Bibliographic information published by the German National Library:

The German National Library lists this publication in the National Bibliography; detailed bibliographic data are available on the Internet at http://dnb.dnb.de .

Imprint:

Print and binding: Books on Demand GmbH, Norderstedt Germany
ISBN: 9783668665446

This book at GRIN:

https://www.grin.com/document/417008

Anastasia Kokkinou

Precambrian Tectonics. Tectonics throughout the Hadean and Archean Eon

GRIN Verlag

Precambrian Tectonics

STRUCTURAL GEOLOGY ASSIGNMENT

Kokkinou Anastasia
WiSe 2014-2015

Kokkinou Anastasia, M.Sc. Resources & Energy

PRECAMBRIAN TECTONICS
STRUCTURAL GEOLOGY ASSIGNMENT

TABLE OF CONTENTS

ABSTRACT

This assignment exhibits the period since Earth has formed until boundary conditions and tectonics started to stabilize. This period is called Precambrian and consists the 7/8 of Earth's history. Precambrian lasts 3,9 Ga and is separated into three Eons, Hadean, Archean and Proterozoic with respect to time occurrence.

Earth loses heat through time due to gradually decrease in decay of radiogenic isotopes which are dispersed into the core, mantle and continental crust. Therefore, Earth composition and tectonics distinctly change. Tectonics generally change from vertical patterns to horizontal and the continental crust forms until the end of the Precambrian to the 82% of the overall crust, covering Earth now. Moreover, different movements of Supercontinents which fragmented, dispersed and collided where studied through geological, paleomagnetic and paleoclimatic studies.

INTRODUCTION

First Cratons

On the basis of isotopic data study, Condie (2005b) concluded that 39% of the continental crust formed during the Archean Eon, 31% during the Early Proterozoic and 12% during the Middle-Late Proterozoic. This means that 82% of the overall Earth's continental crust was formed during Precambrian Period. This is prospective, considering that this period lasts 3,9 billion years, about 7/8 of total Earth's age (Fig. 1). The formation of the first cratonic nucleii marks the transition from the Early Earth's, Magma Ocean, geology when conditions were much hotter and energetic to a state where crustal preservation until present became possible. Main characteristic is a high velocity mantle root, stiff, chemically buoyant and highly resistant which extends to depths at least 200km (King, 2005). The chemical consistence of this mantle root contributed to the long-term survival of the Archean continental lithosphere (Carlson et. Al., 2005).

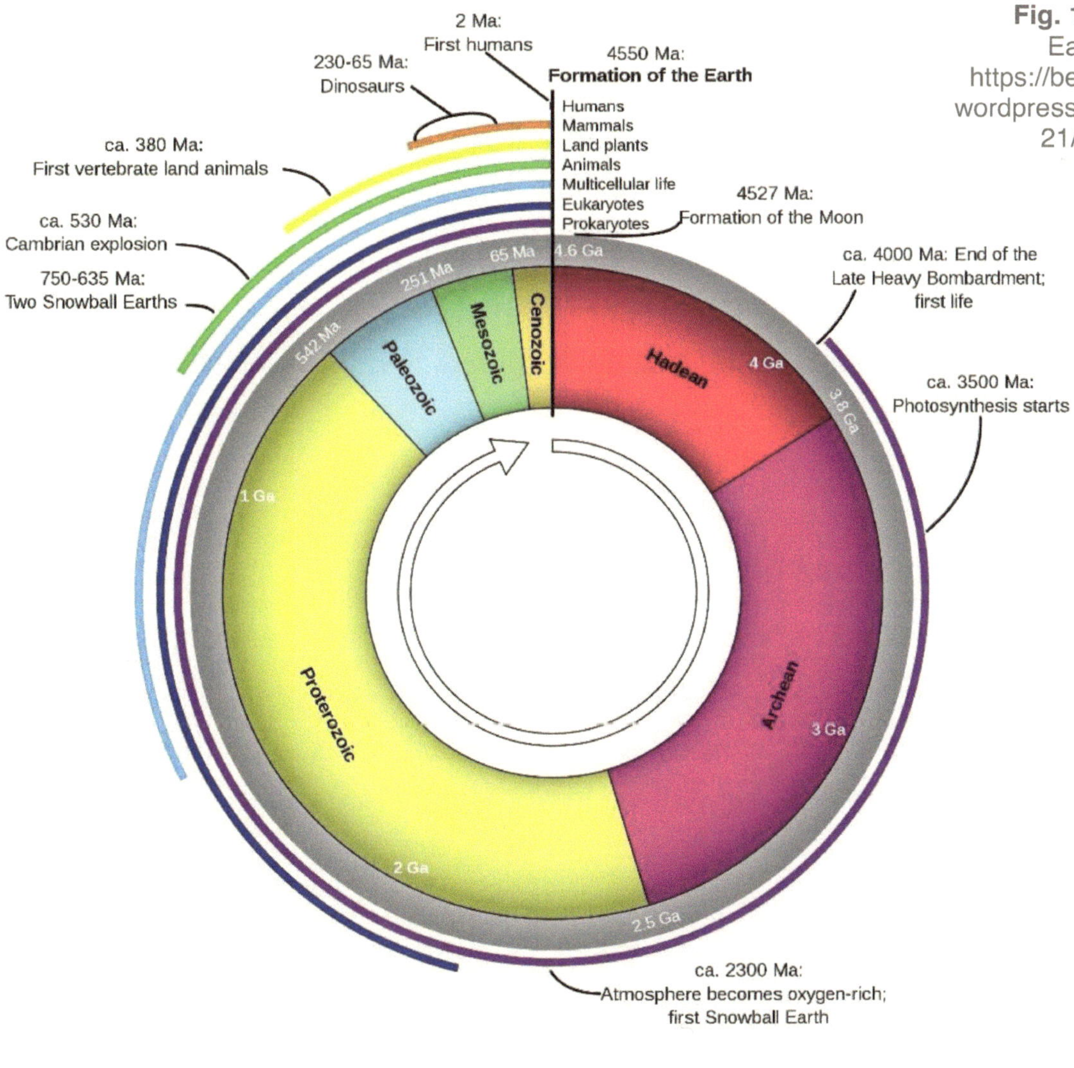

Fig. 1 Summarizing Earth's Timeline. https://bellavitabyjamie.wordpress.com/2012/07/21/timeline/earth-timeline/

The age of the oldest rocks found was originally assumed to have been 4 billion years old: the Acasta gneisses of the Slave Craton in Northwestern Canada (Bowring & Williams, 1999). However, >4,4 detrital Zircon minerals found in the Yilgarn Craton of Western Australia (Wilde et. Al., 2001) and the Greenstone of Nuvvuagittuq Fold Belt which is located in the Minto Block, part of the NE Superior Province, Canada (David et al, 2009 GSA Bulletin) suggest that some continental crust may have formed as early as 4,4-4,5 Ga but still this interpretation is controversial and not yet determined. At the Archean – Proterozoic Boundary there is a gradual infer in tectonic patron observed which is caused by secular Earth cooling due to reduction of radiogenic isotopes. Central debating issues of Precambrian are how tectonics were before this period and if they were likely to the present, uniformitarian models and non-uniformitarian. Towards this determination considering the nature of Precambrian tectonic processes, three approaches have been adopted (Kröner, 1981; Cawood et al., 2006). First, a strictly uniformitarian model which supports that the same exact mechanisms of plate tectonics applied during Phanerozoic Period were also applying during Precambrian Period. This model best fits to interpret Proterozoic Orogenic Belts. Moreover it has been partly applied to Archean Cratons and also to some Hadean belts based on latest studies (T. Rushmer et. Al., 2013; 2014). Second, a moderate uniformitarian model was postulated, in which plate tectonic processes in the Precambrian were somewhat different from present justified by the physical conditions affecting the crust and mantle changes throughout geologic time. This model has been used in studies of both Archean and Early Proterozoic geology. Third model supports no similar tectonics to modern. This latter, non-uniformitarian approach most often is applied to the Hadean and Early - Middle Archean.

Precambrian Heat Flow

Heat Flow is the triggering factor for all major changes over Earth. The majority of terrestrial heat production comes from the decay of radioactive isotopes dispersed throughout the core, mantle, and continental crust which are not renewable. Consistent destruction and reduction of radiogenic isotopes over Earth lead to lower energetic states implying heat flow in the past must have been considerably greater. For an Earth model with a K/U ratio derived from measurements of crustal rocks, the heat flow in the crust at 4,0 Ga would have been three times greater than at the present day and at 2,5 Ga about two times the present value (Mareschal & Jaupart, 2006).

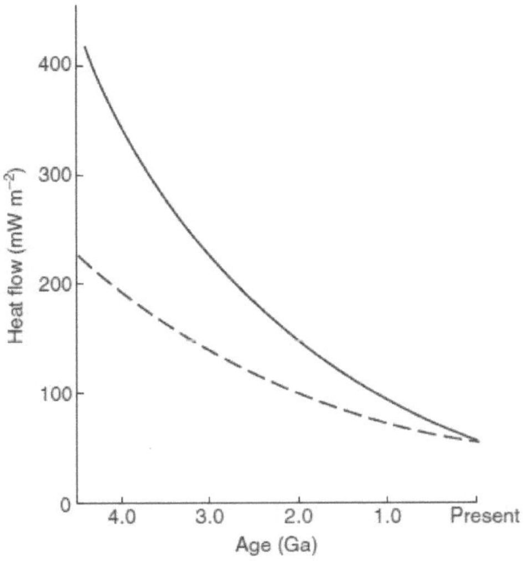

Fig. 2 Variation of surface heat flow with time. Solid line, based on a chondritic model; dashed line, based on a K/U ratio derived from crustal rocks (redrawn from McKenzie & Weiss, 1975, with permission from Blackwell Publishing). Kearey et. Al..2008. Global Tectonics. Ch. 11, p.348

If the heat loss mostly occurred by the relatively inefficient mechanism of conduction then the lithosphere would have been warmer in present time. However, if the main mechanism of heat loss was convection beneath oceanic lithosphere, which is very effective at dissipating heat, then the continental lithosphere need not have been much hotter during Precambrian (Lenardic, 1998). Clarifying these aspects of the Archean thermal regime is essential in order to reconstruct tectonic processes in the ancient Earth and to assess whether they were different than they are today. Geologic evidence from many of the cratons, including an abundance of high temperature/low pressure metamorphic mineral assemblages and the intrusion of large volumes of granitoids, suggest relatively high (500–700 or 800°C) temperatures in the crust during Archean times, roughly similar to those which occur presently in regions of elevated geotherms. By contrast, geophysical surveys and isotopic studies of mantle nodules suggest that the cratonic mantle is strong and cool and that the geotherm has been relatively low since the Archean (Pearson et al., 2002; Carlson et. al., 2005). But it is determined to have been much thicker than current lithosphere through heat determining studies. According to Sleep (2003) model, a chemically buoyant layer of lithosphere forms a highly resistant lid above the convecting mantle, allowing it to maintain nearly constant thickness over time. These considerations illustrate how the formation and long-term survival of the cool mantle roots beneath the cratons has helped geoscientists constrain the mechanisms of heat transfer during Precambrian times.

Hargraves (1986) concluded that heat loss through the oceanic lithosphere is proportional to the cube root of the total length of the mid-ocean ridge. Assuming a nonexpanding Earth, the increased rate of plate production implies a similar increase in plate subduction rate. These computations suggest that some form of plate tectonics was taking place during the Precambrian at a much greater rate than today, main Archean characteristic. The fast rates suggest an image of the solid surface of the early Earth where the lithosphere was broken up into many small plates that contrasts with the relatively few large plates that exist presently.

Hadean Eon 4,5 – 3,9 Ga

Beginning with the oldest Earth's history, Hadean Eon is the period Earth suffered consistent meteorite bombardments. Heat flow was a lot greater than modern times and was assumed to have been covered by magma oceans. However, there are no sufficient evidence to determine actual current conditions. Based on latest studies (T. Rushmer et. Al., 2013; 2014) there are serious implications of some kind of limited horizontal tectonics. Moreover other scientists don't even consider the existence of Hadean Eon as a separate period different from Archean period and place Archean lower boundary to the age of 4,5 Ga. Interpretations for the Hadean Eon are keenly controversial.

Earth's oldest rocks studies

TTGs Studies

The Earth's oldest continental crust is composed primarily of granitic rocks known as tonalite-trondhjemite-granodiorite (TTG) (Jahn et al., 1981; Rudnick, 1995; Condie, 2005). These granites are classically associated with belts of metamorphosed volcanic rocks and sediments known as greenstones. Although it is generally believed that TTG magmas were produced by partial melting of a basaltic crust that predated them (Drummond and Defant, 1990; Martin, 1994; Rapp et al., 2003; Condie, 2005), the tectonic environment in which this occurred is not known. Furthermore, whether there is a direct genetic relationship between TTGs and their immediate greenstone host rocks is also problematic (Martin,

1987, 1994). These uncertainties have broader implications for studies of early Earth processes, including the question of when plate tectonics first became active. T. Rushmer et. Al. (2013) have approached these issues by conducting partial melting experiments on two greenstones from one of the oldest known greenstone belts, the Nuvvuagittuq complex of northern Quebec (O'Neil et al., 2008, 2011). The Nuvvuagittuq greenstones include two stratigraphically and compositionally distinct groups (see O'Neil et al., 2011): these are (1) an incompatible element–depleted tholeiitic group, and (2) an overlying calc-alkaline group that is moderately enriched in incompatibles. Their objective was to test for a direct genetic link between the second overlying calc-alkaline group of Nuvvuagittuq greenstones and a closely associated but younger tonalite of the TTG series. TTGs are experimentally defined to have generated at 3,8 Ga by melting of 3,8 – 4,3 Ga mafic, arc-like, material as shown by expanded ^{142}Nd isotope data set and Sm/Nd systematics (O'Neil et al., 2008). At pressures of 0.5 to 2.0 GPa orthopyroxene ± olivine are the only near-liquidus phases, consistent with results for Tertiary boninites (Fig. 3). Under subliquidus conditions, at moderate to high degrees of partial-melting (34-47 %), melt coexists with clinopyroxene ± orthopyroxene ± plagioclase ± garnet. The melts are compositionally similar to TTGs, including the 3.66 Ga tonalite that encloses the Nuvvuagittuq greenstone belt (T. Rushmer et. Al., 2013).

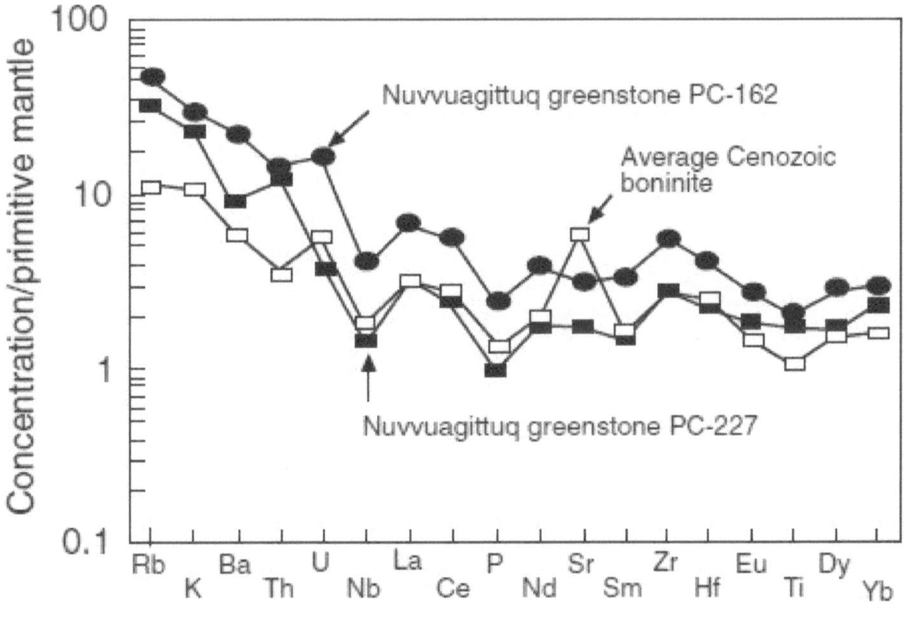

Fig. 3 Incompatible element concentrations in Nuvvuagittuq greenstones PC-227 and PC-162, and average Cenozoic boninite (using data from Cameron et al., 1983). Data are normalized to primitive mantle (Sun and McDonough, 1989) concentrations. T. Rushmer et. al., 2013. Hadean greenstones from the Nuvvuagittuq fold belt and the origin of the Earth's early continental crust

Thus in contrast to many previous proposals (e.g., Drummond and Defant, 1990; Foley et al., 2002), the distinctive compositional characteristics of TTGs (including low Nb, Ta, and Ti and high Rb, K, and Ba) may be the result of chemical fractionation that preceded TTG magma formation (T. Rushmer et. Al., 2013). This possibility is consistent with a number of other observations. One is that not all basalts can produce TTG-like melts. In fact, of the many experiments conducted on basaltic starting materials, very few have produced melts that genuinely resemble TTGs (T. Rushmer et. Al, 2013). It has also been shown that the concentrations of most incompatible elements in Archean TTGs are too high for them to have been produced by the partial melting of the K-poor tholeiites that predominate in most greenstone terrains, but instead require crustal sources that were already enriched in these components (Smithies et al., 2009). The fact of enriched Hadean crust predating the earliest known continental crust has been independently verified by studies of Pb, Nd, and Hf isotopes in Paleoarchean rocks and zircons (Andreasen and Sharma, 2009; Blichert-Toft and Albarede, 2008; Tessalina et al., 2010). Taken together, these observations demonstrate that many of the compositional characteristics of the early continental crust must have been inherited from previous cycles of mantle depletion and subsequent reenrichment (T. Rushmer et. Al., 2013).

However, there is sufficient evidence to suggest that, as in the modern Earth, the crust of the Hadean Earth was organized into a number of different tectonic environments and that one of these gave rise to the first continental crust. This is consistent with some form of horizontal tectonics and difficult to reconcile with models based purely on plume-driven tectonics (T. Rushmer et. Al., 2013). An alternative explanation based on isotope and geochronologic data is that the Nuvvuagittuq lavas were generated ca. 3.8 Ga, and that the variations in ^{142}Nd isotopic compositions were inherited from primordial differentiation of the Earth's mantle (Roth et al., 2013; Guitreau et al., 2013).

Hadean detrital Zircon

Numerous studies have utilized isotopic systems in both whole rocks and detrital minerals to investigate the geodynamics of the early Earth. For example, several groups have used the compositions of Hadean detrital zircons from the Jack Hills in Australia to argue that the crust of the early Earth was not so different from today (Mojzsis et al., 2001; Wilde et al., 2001; Harrison, 2009), whereas others have suggested that the early crust was more basaltic (e.g., Kemp et al., 2010). Some data sets have been further used to infer that subduction did not commence until 3 Ga (e.g., Dhuime et al., 2012) or even much later (Stern, 2005). Unfortunately, old rocks are rare and detrital minerals lack important geological context. T. Rushmer et. Al. (2015?) suggest that this geochemical stratigraphy might provide a more robust test of ancient tectonic settings than individual chemical or isotopic signatures in rocks or detrital minerals.

On a global basis, there are some long noted and intriguing similarities between (non-komatiitic) Archean mafic rocks and those found at present- day subduction zones. For fluid-immobile elements, these include elevated concentrations of Th and Zr relative to Nb, Ti, and Yb (Fig. 4), effectively the negative Nb and Ti anomalies that are characteristic of arc lavas and continental crust in general. There are also differences between the geochemical patterns of modern and Archean mafic rocks. For example, whereas modern mafic rocks show a clearly bimodal distribution between arc and non-arc settings, the Archean mafic rocks spread across these two fields (Fig. 4) (Moyen, 2013). Thus, geochemists have been appropriately prudent in assigning individual geochemical signatures to tectonic settings for Archean rocks (Pearce, 2008).

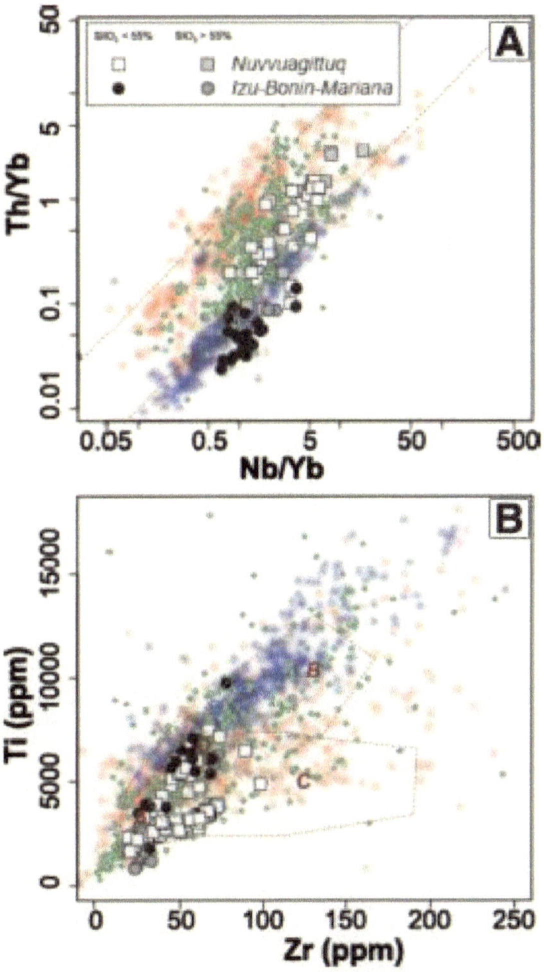

Fig. 4 Trace element discrimination diagrams showing comparison of modern arc basalts (red cloud) and mid-oceanic ridge basalts (MORB, blue cloud), non-komatiitic Archean basalts from GEOROC (green diamonds; Geochemistry of Rocks of the Oceans and Continents; http://georoc.mpchmainz.gwdg.de/georoc/), and mafic rocks from proto Izu-Bonin-Mariana arc (black circles) and Nuvvuagittuq (Quebec, Canada) supracrustal belt (white squares). Samples with SiO2 > 55 wt% are shown in gray. A: Th/Yb versus Nb/Yb diagram from Pearce (2008). B: Ti (ppm) versus Zr diagram from Pearce and Cann (1973). Field A is island arc tholeiite (IAT); field B is MORB; field C is calc-alkaline basalt (CAB); f eld D is ambiguous(intersection of previous three fields). T. Rushmer et. Al., 2015?. Heading down early on? Start of subduction on Earth.

The initiation of modern-day subduction has long been thought to be marked by eruption of distinctive lavas, called boninites, which are characterized by depletion in relatively compatible elements such as Ti. However, recent submersible exploration in the Izu-Bonin-Mariana forearc (Reagan et al., 2010; Ishizuka et al., 2011) found that the lowermost rocks are actually similar to mid-oceanic ridge basalts (MORBs). These are overlain successively by boninites associated with hydrothermal Fe-oxide and sulfide deposits (Ishizuka et al., 2008) and then by rocks of arc tholeiitic and calc-alkaline affinity that are typical of modern subduction systems (Fig. 5). The key outcome is that subduction initiation can now be identified by a geochemical stratigraphy (Stern et al., 2012), rather than the occurrence of one single lava type (e.g., boninites) or a selected trace element signature (such as negative Nb anomalies).

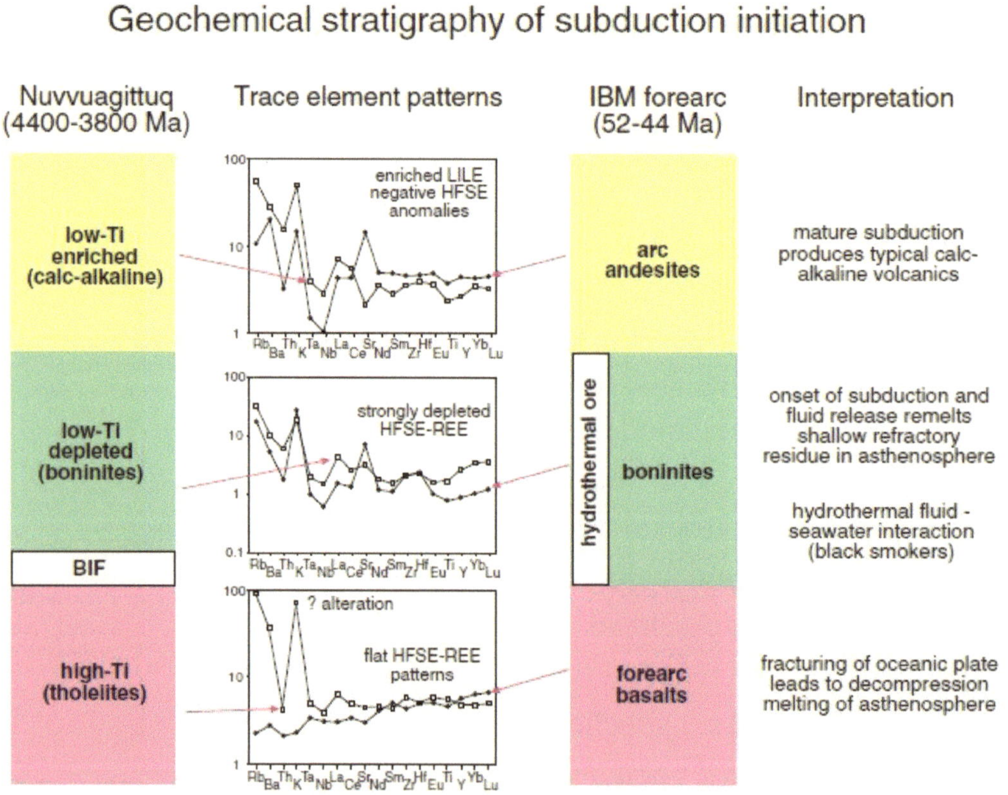

Fig. 5 Despite possible 4,4 Ga difference in age, stratigraphic succession and trace element composition of old Earth rocks (O'Neil et al., 2011, 2012) are strikingly similar to those formed during inception of Izu-Bonin-Mariana (IBM) arc today (Reagan et al., 2010). Geochemical stratigraphy consists of four successive units that reflect transition from initial fracturing of an oceanic plate through commencement of fluid release, melting, and finally mature subduction. Trace element patterns are individual analyses: Nuvvuagittuq (Quebec, Canada) samples are PC132 (high Ti), PC432 (low Ti), and PC149 (low Ti enriched) from O'Neil et al. (2011); IBM samples are 974-R9 g (fore-arc basalt), 974-R4 (boninite), and GU9 (calc-alkaline) from Reagan et al. (2010) and Woodhead (1989). BIF—banded iron formation; LILE—large ion lithophile elements; HFSE—high field strength elements; REE—rare earth elements. Rushmer et. Al., 2015?. Heading down early on? Start of subduction on Earth.

The high-Ti tholeiites have flat Rare Earth Element (REE) and High Field Strength Element (HFSE) patterns that typify basalts generated from a relatively undepleted mantle. In contrast, the overlying low-Ti mafic rocks have high MgO and SiO2, are highly depleted in REEs and HFSEs, and have concave-upward REE patterns that are typical of boninites generated from previously depleted mantle. In both cases, they could represent hydrothermal activity associated with early extension, and perhaps the first introduction of subducted fluids (those associated with generation of the low-Ti lavas) (T. Rushmer et. Al., 2015?).

Some differences, such as the greater abundance of mafic sills and ultramafic rocks in the Nuvvuagittuq stratigraphy (Fig. 5), could indicate that there were local differences in magma generation during rifting and subduction.

Hadean tectonic models

There have been several geotectonical models associated either with the TTGs formation or the detrital zircon isotope according to above mentioned studies.

TTGs

<u>Vertical Tectonic approach:</u>

- Mantle melting of mafic material derived from the lower mantle as shown by recent synthesis of TTG host rock and zircon Hf isotopes,
- Then evolving at shallower levels, or by direct partial melting, as suggested by experimental melting studies.
- Process remains similar through time.

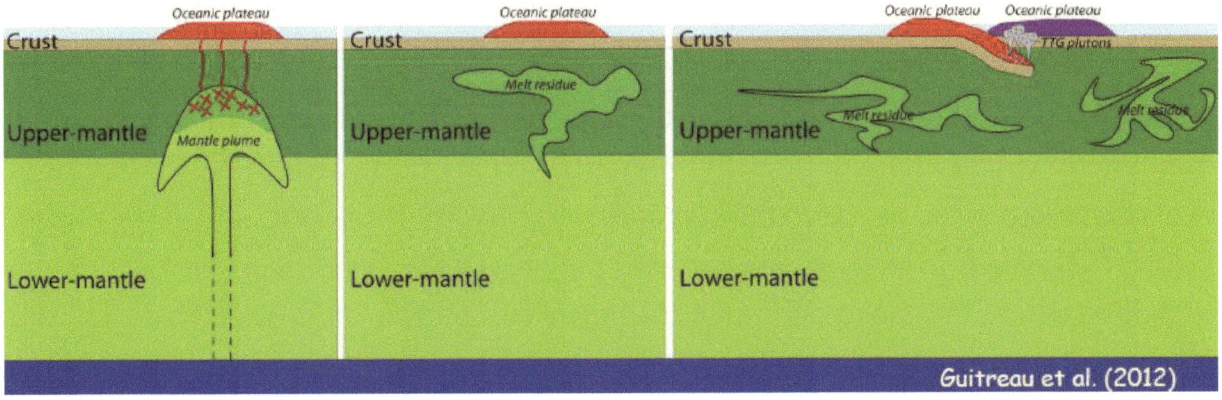

Fig. 6 TTGs formation derived from mantle plume source. Vertical tectonic Model. (Guitreau et. Al., 2012) Rushmer et. Al., 2014. Onset of subduction and the genesis of TTGs (tonalities-trondhjemites-granodiorites) in the in the earliest Earth. (Powerpoint presentation)

Horizontal Tectonic approach:

The initiation of Hadean Eon subduction procedures was introduced by several scientists (Albarde, 1998, Boher et al., 1992; Condie 2005; Kerrich & Polat, 2006; Bedard, 2006; Polat, 2012; Rushmer et. Al., 2013).

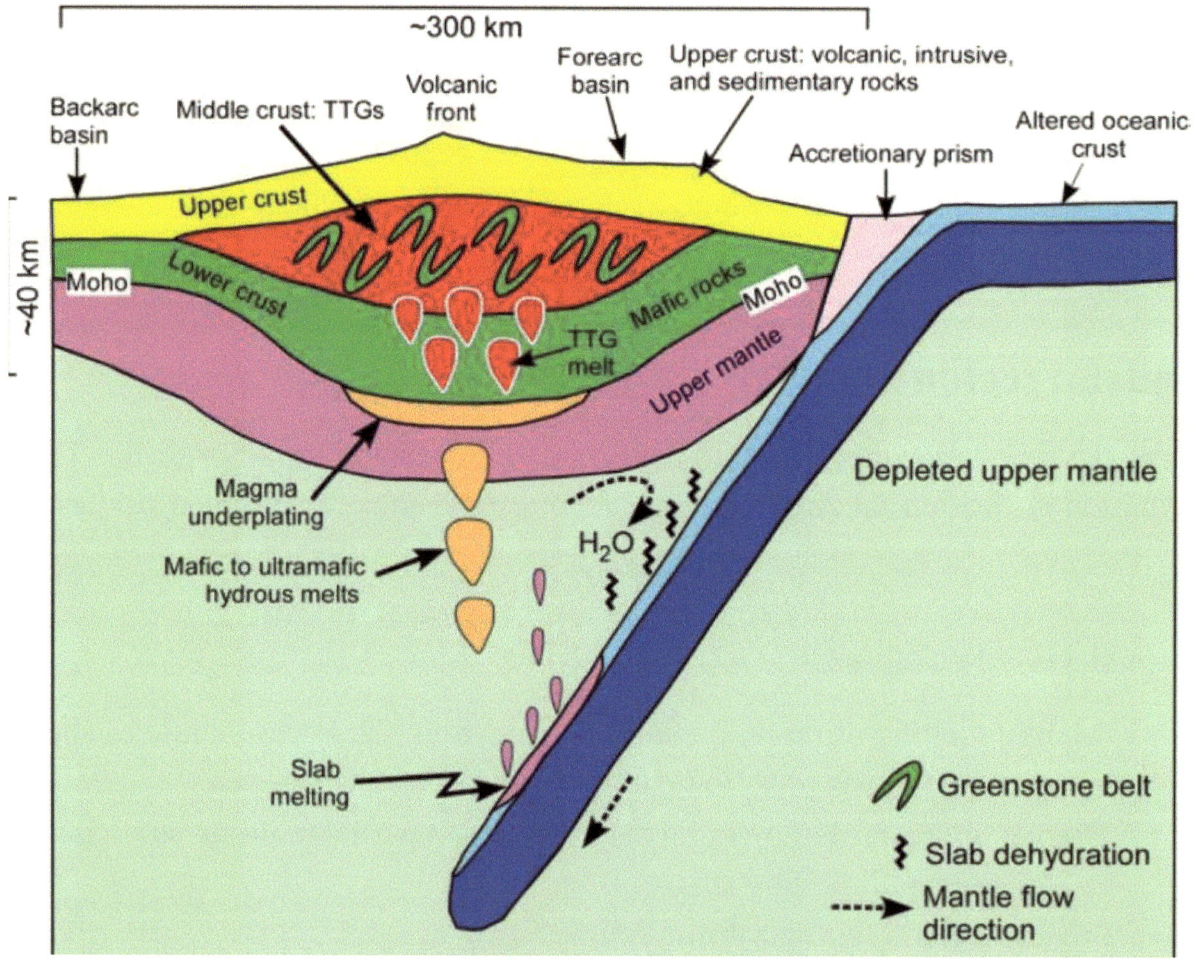

Fig. 7 TTGs formation during oceanic lithospheric subduction. Horizontal tectonic Model. Rushmer et. Al., 2014. Onset of subduction and the genesis of TTGs (tonalities-trondhjemites-granodiorites) in the in the earliest Earth. (Powerpoint presentation)

Main point of this approach is the role of oceanic plateaus and plumes and plate interaction (subduction zone geodynamics as today) during Early Earth geology. The subducted plate is dehydrating and partially melting (Fig. 7). The mafic-ultramafic composition melts ascent until they reach the upper mantle- lower crust boundary where the magma is unterplating. After partial melting of this material TTGs are formed and intrude the crustal formations.

Zircon:

This model focuses on Zircon detrital isotope initiation and is mainly based on isotopic data determination from Jack Hills zircon study by Kemp et al., (2010). Earth composition is conducted by nascent hydrosphere, steam atmosphere and magma oceans (Fig. 8). Upper magma layer is altered and hydrated KREEPy rind with $\delta^{18}O$ 5-7‰ underlined by thin KREEPy basaltic crust (dry $\delta^{18}O$ 5,3‰). KREEP is material having very high concentrations of incompatible elements; its name is the acronym for the incompatibles K, Rare-Earth-Elements (REE) and P. Hadean depleted mantle or magma-ocean cumulate begins ascending into upper KREEPy layers and erupts above them. The Zircon of Jack Hills is connected with the solidification and crystallization of this magma ocean (KREEPy layer) and a prolonged period of partial melting.

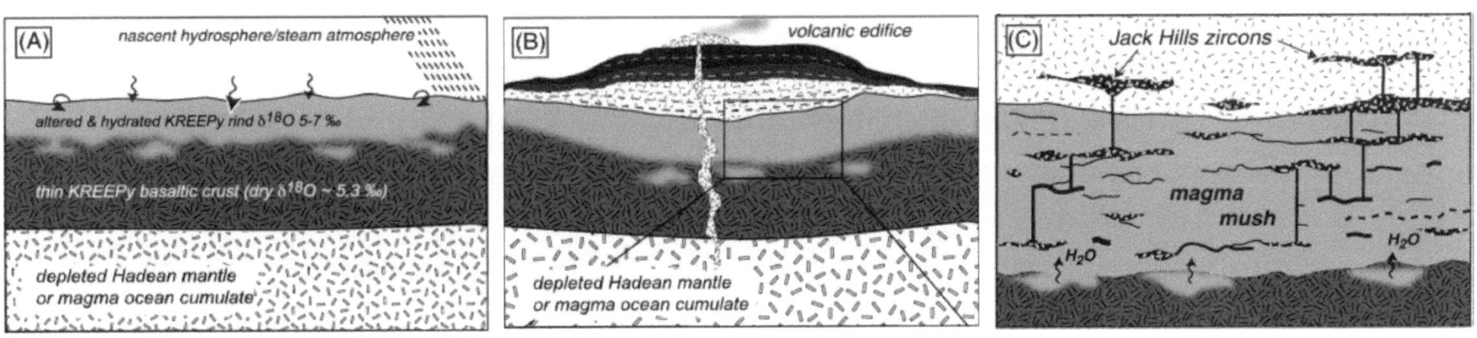

Fig. 8 Hadean Zircon initiation: From Jack Hills zircon study (Kemp et al., 2010), solidification and crystallization of a magma ocean (KREEPy layer) and a prolonged period of partial melting. Rushmer et. Al., 2014. Onset of subduction and the genesis of TTGs (tonalities-trondhjemites-granodiorites) in the in the earliest Earth. (Powerpoint presentation)

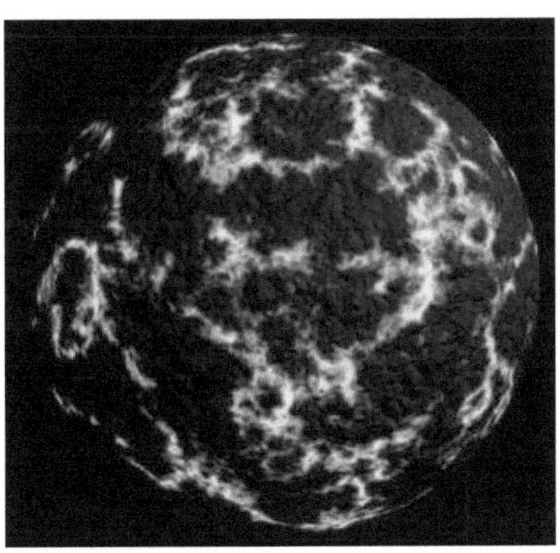

Concluding for Hadean Eon:

Definite Hadean (primoridal) magma differentiation is observed. Non horizontal tectonics until Archean (Kerrich & Polat, 2006; Shirey et al.2008; Polat, 2012) is supported. Maybe some type of plates and some very limited horizontal tectonic movements' implications (Turner et. Al., 2015?) (Fig. 9).

Fig. 9 Hadean Earth's image reconstruction. Tracy Rushmer. 2014. Onset of subduction and the genesis of TTGs (tonalities-trondhjemites-granodiorites) in the in the earliest Earth. (Powerpoint presentation)

Archean Eon 3,9 – 2,5 Ga

The Archean Eon is mainly characterized by vigorous production and destruction of oceanic lithosphere. This procedures led to the production of voluminous calc-alkaline melt and composing of the Archean tonalitic rocks. Also extensive rifting in marginal basins took place at this period, giving rise to multiple greenstone belts. There are no rigid plates or stable cratons at this period. On the other hand Archean lithospheric mantle roots are preserved until present time at several places over the Earth. What contributed to the preservation of Archean lithospheric roots was the chemically buoyant material which was stiff and resistant.

General characteristics of cratonic mantle lithosphere

Archean cratonic mantle lithosphere is characterized by high seismic velocity to depths of at least 200km in contrast to Proterozoic belts and by lowest surface heat flow of any other region on Earth. The base of the continental lithosphere is not well defined by seismological data but is suggested to be ~250km deep in undisturbed cratonic areas and ~180km beneath Phanerozoic rifts and orogens (O'Reilly et. Al., 2001). Most of it is confirmed through isotopic Re-Os studies to have remained thermally and mechanically stable over the past 3-2 Ga. This indicates that roots of the Archean cratons are cool, strong and compositional distinct from the surrounding mantle (Kearey at. Al., 2008).

In addition to being cool and strong, studies of mantle xenoliths indicate that the Archean mantle roots are chemically buoyant and highly depleted in incompatible elements (O'Reilly et al., 2001; Pearson et al., 2002). Both buoyancy and chemical depletion are achieved simultaneously by partial melting and melt extraction, which, in the case of the mantle lithosphere, has left behind a residue composed of Mg-rich harzburgites, lherzolites, and peridotite (O'Reilly et al., 2001). These characteristics have resulted in the mechanical and thermal stability of the cratons for up to three billion years.

General Geology of Archean Cratons

Archean cratons are composed by two groups of metamorphic rocks which are distinguished by their metamorphic grade: a) Greenstone belts and b) High grade gneiss terrains (Windley, 1981). Both groups are intruded by large volumes of granitoids, formed by compositionally distinctive group known as TTG suites (Tonalite-Trondjemite-Granodiorite).

Greenstones

Archean greenstone belts consist of metavolcanic and metasedimentary rocks that exhibit a low pressure (200-500 MPa), low temperature (350-500° C) regional metamorphism of greenschist facies. Their lower group is composed of tholeitic and komatiitic lavas. Komatiites are varieties of Mg-rich basalt and ultramafic lava that occur almost exclusively in Archean crust which were used to infer melting temperatures that are higher than those of modern basaltic magmas (Kearey et. Al., 2008). The central group contains intermediate and felsic volcanic rocks whose trace and rare earth elements are similar to those found in some island arc rocks. The upper group is composed of clastic sediments, such as graywackes, sandstones, conglomerates, and banded iron formations (BIFs).

High-grade gneiss terrains

High gneiss terrains form the majority of the area of Archean cratons and are characterized by low pressure, high temperature (>500° C) regional metamorphism of the amphibolite or granulite facies. They are found highly deformed and may have formed either contemporaneously with, structural bellow, or adjacent to the low-grade greenstone belts (Percival et. Al., 1997).

Formation of the Archean mantle lithosphere

The distinctive composition and physical properties of the stiff, buoyant mantle roots beneath the cratons result from the chemical depletion and extraction of melts from the primitive mantle. These two processes lowered the density and increased the viscosity of the residue left over from partial melting and resulted in a keel that consists mostly of high-Mg olivine and high-Mg orthopyroxene

(O'Reilly et al., 2001; Arndt et al., 2002). The distinctive composition is related to unusually high degrees (30–40%) of mantle melting over a range (4–10 GPa) of mantle pressures (Pearson et al., 2002; Arndt et al., 2002). High-degree partial melting of mantle peridotite produces magma of komatiitic composition and a solid residue that is very similar to the composition of Archean lithospheric mantle (Herzberg, 1999; Arndt et al., 2002). Although high-degree partial melting undoubtedly occurred, this process alone cannot explain the origin of the Archean mantle lithosphere. The main reason for this conclusion is that the abundance of komatiite found in the Archean crustal record is much too low to balance the amount of highly depleted peridotite found in the cratonic mantle (Carlson et al., 2005). According to Arndt et. Al. (2002) reconstructions, there are three possible approaches that could have resulted in the mechanical segregation and accumulation of such a material near the Earth's surface (Fig. 10).

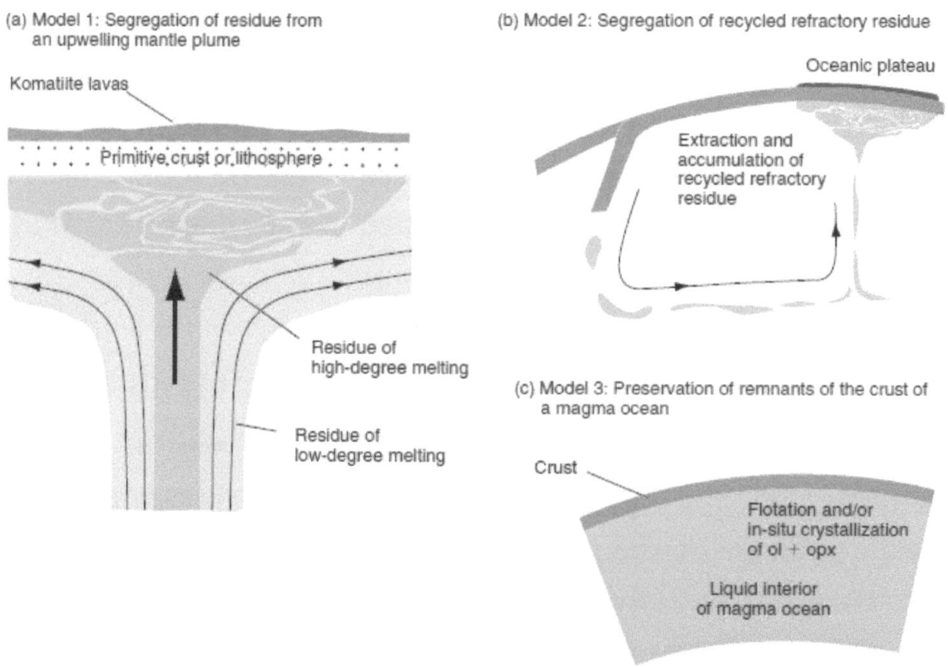

Fig. 10 (a–c) Three possible mechanisms that could allow the segregation and accumulation of high-Mg olivine and orthopyroxene near the surface of the Earth (after Arndt et al., 2002, with permission from the Geological Society of London). Kearey et. Al.. 2008. Global Tectonics. Ch. 11 Precambrian Tectonics, p. 352

a) Upwelling buoyant residue in the core of a mantle plume could have separated from the cooler, denser exterior and could have accumulated during ascend. In this model, melting begins at high pressure (~200km depth) and continues to shallow depths, by which point melt volumes are high and the dense residue of early, low degree melting is swept away by mantle flow (Fig. 10 (a)).
b) In this approach, buoyant residue could have segregated slowly as material was transported down through subduction zones and recycled through the mantle in convection cells (Fig. 10 (b)).

c) Based on this model, some subcontinental lithosphere could be the remnants of an initial crust, crystalized in a Hadean magma ocean that formed during the final stages of Earth accretion (Fig. 10 (c)).

By applying a range of criteria some geologic studies have made compelling cases that ancient mantle plumes played a key role in the evolution of Archean lithosphere (Tomlinson & Condie, 2001; Ernst et al., 2005). Nevertheless, it's important to recognize that Archean mantle roots probably resulted from more than one tectonic environment and that no single setting or event is applicable to all cases (Kearey et. Al., 2008).

Komatiites, part of the greenstone belts series, are characterized by high magnesium contents and high degrees of melting (1400-1600º C). Therefore they were associated with the production of Archean lithospheric mantle roots and were used to infer current tectonic model. Komatiitic and tholeitic lavas loosely represent the Archean equivalent of modern mid-ocean ridge basalts, considering mantle was hotter by some amount than the modern mantle, which undoubtedly influenced the compositions, source depths and patterns of the volcanism (Nisbet et. Al., 1993).

If the Komatiites were produced by the melting of hydrous mantle, then the depth of melting to produce this type of residue could have been relatively shallow, as in subduction zones, and the Archean mantle temperature needed only be slightly hotter than present. This would result to the formation and thickening of highly depleted mantle lithosphere which has later been incorporated into the cratonic mantle below the continent (Fig. 11).

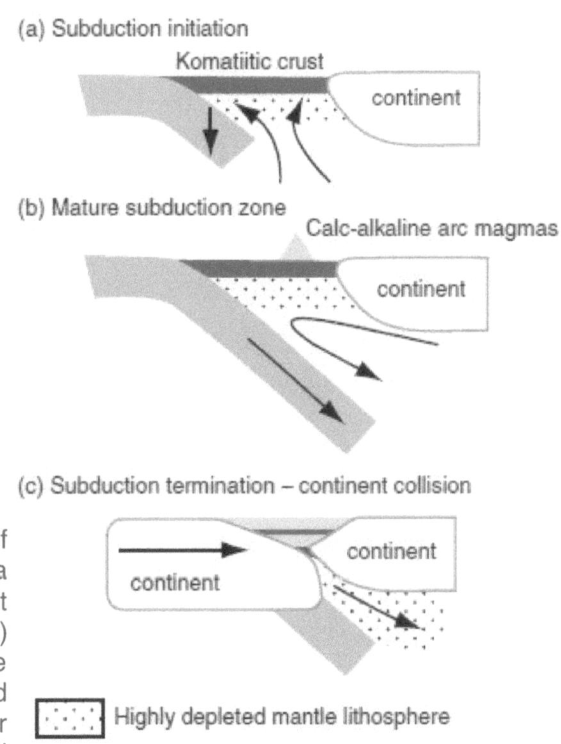

Fig. 11 Conceptual model for the generation of komatiites and cratonic mantle by partial melting in a subduction zone (after Parman et al., 2004. Copyright © 2004 Geological Society of South Africa). (a–c) Three possible mechanisms that could allow the segregation and accumulation of high-Mg olivine and orthopyroxene near the surface of the Earth (after Arndt et al., 2002, with permission from the Geological Society of London). Kearey et. Al.. 2008. Global Tectonics. Ch. 11 Precambrian Tectonics, p. 353

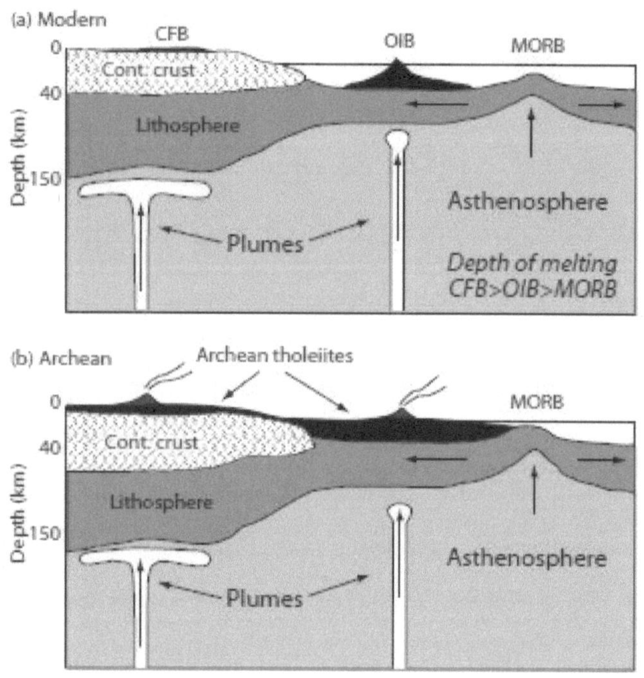

Fig. 12 Model of komatiitic and tholeiitic basalt formation involving mantle plumes (after Arndt et al., 1997, by permission of Oxford University Press). Model shows the influence of lithospheric thickness on depth of melting where CFB is continental flood basalt, OIB oceanic island basalt, and MORB mid-ocean ridge basalt. Kearey et. Al.. 2008. Global Tectonics. Ch. 11 Precambrian Tectonics, p. 353

Alternatively if the source rock of Komatiites were dry then the high ambient temperatures in the Archean mantle would have caused melting to begin at depths, much greater than those occur in subduction zones. Possibly melting would initiate in upwelling mantle plumes or at unusually deep levels (~200km) beneath mid-ocean ridges (Fig. 12). The greater depths of melting can give rise to large volumes of basalt and oceanic crust, which would be much thicker (20-40km) (Fig. 12 (b)) than modern (Fig. 12 (a)) (Bickle et. Al., 1994).

Using those two models for the formation of Archean mantle lithosphere, the differences between the modern and ancient rocks can be explained by variations in the depth of melting and on the effects of the much thicker Archean overriding lithosphere. However the fact of a much thicker crust which might have inhibited subduction, is more controversial. As an alternative to the subduction model, Zedgers & Van Keken (2001) postulated that TTG suites formed by the removal and sinking of the dense, lower part of thick oceanic plateau. The peeling away, or delamination, of a dense eclogite root results in uplift, extension, and partial melting to produce TTG suite magmas. As the oceanic crust cooled and became thinner, a point was reached in the Late Archean when the entire crust could subduct. At this time hydrothermally altered crust, such as amphibolite, was introduced into subduction zones and led to the widespread formation of TTG suites. This model supports the view that the formation of the earliest continental crust requires subduction and the melting of a hydrous mafic source. No certain model is yet considered to be the one contributing the Archean mantle lithosphere alone. Nevertheless, it is important to recognize that Archean mantle roots resulted from more than one tectonic environment and more than one single setting or event is applicable to all cases.

Formation of the Archean crust

The Archean crust was composed of granite-greenstone belts which display a variety of structures and outcrop patterns. Those belts include large tracks of metamorphosed igneous and sedimentary rocks that are deformed by thrust faults and strike-slip zones. In the weakest deformed areas, the delicate Archean stromatolites and other evidence of early life (Buick, 2001), were found and preserved. The special Dome-and-Keel architecture, which occurs only in Archean, consists of trough-shaped or synclinal keels composed of greenstone that surround ellipsoidal and ovoid-shaped domes composed of gneiss, granitoid and migmatite (Kearey et. Al., 2008). The Eastern Pilbara craton of northwestern Australia is one of the oldest and best preserved examples of a granite-greenstone belt and dome-and-keel province with a history spanning 3.72– 2.83 Ga (Collins et al., 1998; Van Kranendonk et al., 2002, 2007). The craton exposes nine granitoid domes with diameters ranging from 35 km to 120 km. The domes display compositional zonation and variable degrees of deformation. This internal structure indicates that each dome was constructed through the emplacement of many intrusions over hundreds of millions of years and that deformation accompanied the magmatism (Kearey et. Al., 2008).

Dome and Keel Tectonics

Dome and keel architecture is an only Archean crust feature and the origin of this special structure has been the subject of many studies because it is important for understanding the nature of current tectonics. Interpretations are divided into mainly horizontal or vertical displacements, producing this pattern.

Vertical Tectonics:

These type of models describe the diapiric rise of hot granitoid domes as the result of a partial convective overturning of the middle and upper crust.

The process begins with the emplacement of hot TTG suite granitoids into an older greenstone succession (Fig. 13a). Domes are initiated at felsic volcanic centers due to a laterally uneven emplacement of TTG magma. After a hiatus of several tens of millions of years, the emplacement of thick piles of basalt on top of less dense granitoids creates an inverted crustal density profile (Fig. 13b). The magmatism also buries the granitoids to mid-crustal depths where they partially melt due to the buildup of radiogenic heat and, possibly, the advection of heat from mantle plume activity. Thermal softening and a reduction in mid-crustal viscosity facilitates the sinking of the greenstones, which then squeezes out the underlying partial melts into rising, high-amplitude granitoid domes (Fig. 13c). The convective overturning depresses geotherms in the greenstone tracts, resulting in local cooling and the preservation of kyanite-bearing metamorphic rocks, which equilibrate at moderate-low pressures (~600 MPa) and temperatures (500°C). This model explains the formation of the dome-and-keel structure without rigid plates or plate boundary forces and is similar to the sinking or sagduction models proposed for the formation of dome-and-keel structures in the Dharwar craton of India (Chardon et al., 1996).

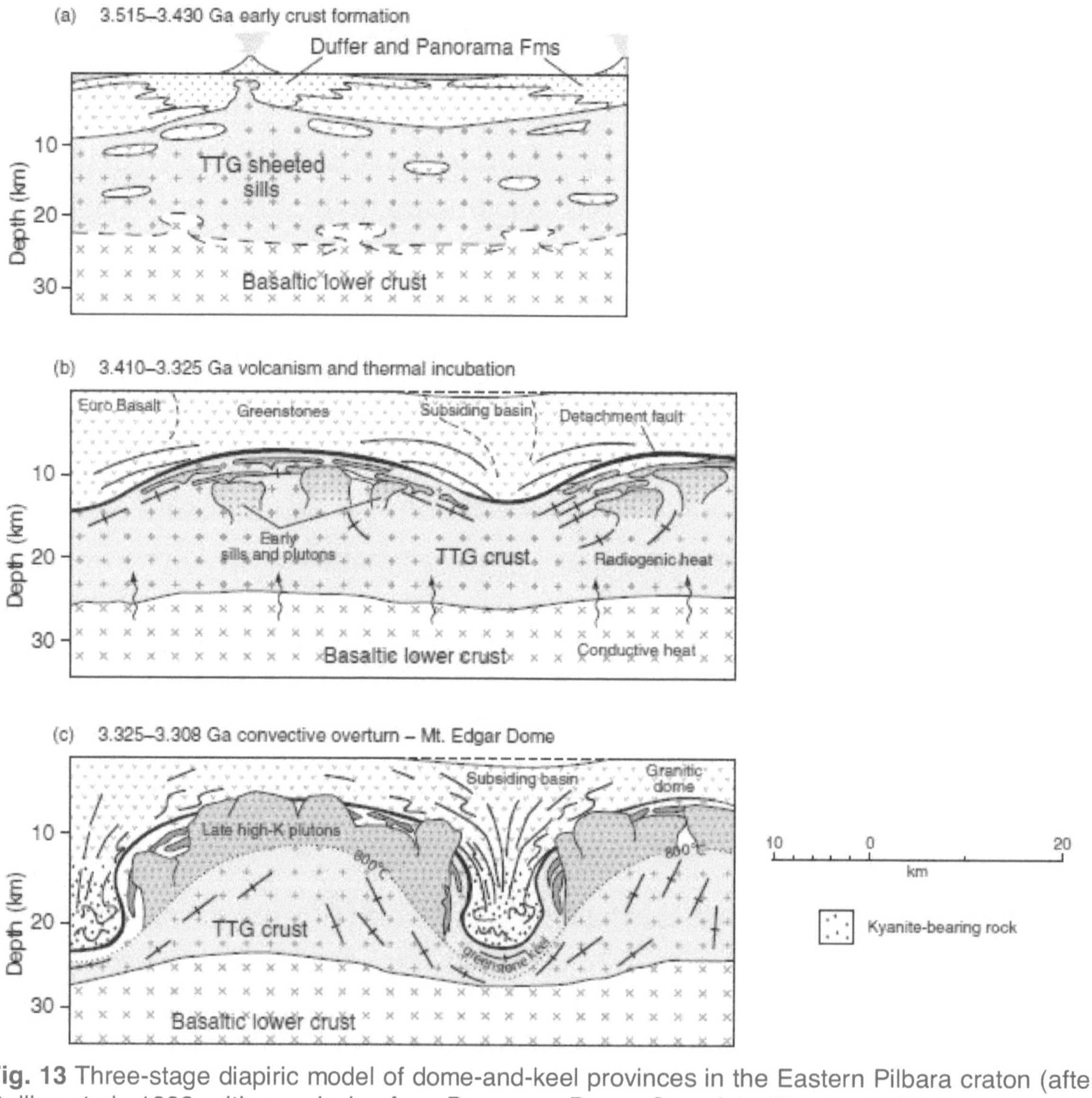

Fig. 13 Three-stage diapiric model of dome-and-keel provinces in the Eastern Pilbara craton (after Collins et al., 1998, with permission from Pergamon Press, Copyright Elsevier 1998; the age-spans of the stages are from Van Kranendonk et al., 2007). Kearey et. Al., 2008. Ch. 11 Precambrian Tectonics, p. 359

<u>Horizontal Tectonics:</u>

Horizontal tectonic models for the Eastern Pilbara propose that the greenstones were affected by one or more periods of horizontal contraction and extension (Blewitt, 2002). In these interpretations, the dome-and-keel architecture results from episodes of Early Archean collision and terrain accretion. Periods of horizontal extension provokes the formation of crustal detachments and the emplacement of the granitoid domes. This model is mainly based on the study of Warrawoona Greenstone Belt (WGB) and Mt. Edgar Dome from Eastern Pilbara Craton (Kloppenburg et. Al., 2001). An initial period of terrain collision and thrusting prior to 3.32 Ga thickens the Early Archean Warrawoona Greenstone Belt and buries granitoid basement to mid-crustal levels where it partially melts. Partial melting facilitates the extensional collapse of the thickened crust, forming detachment faults (i.e. Mt Edgar shear Zone, Fig. 14a). The density inversion

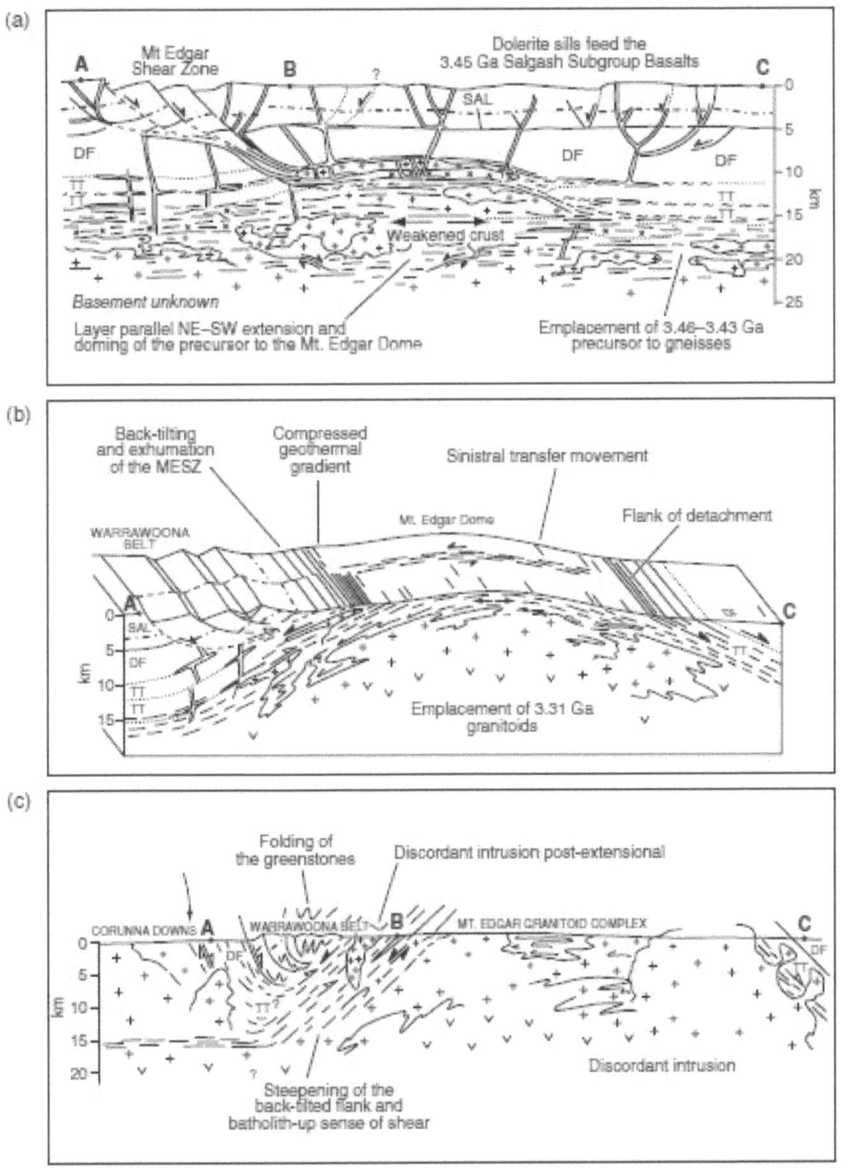

Fig. 14 Cartoon summarizing the tectonic and magmatic development of the Warrawoona Greenstone Belt (WGB) and Mt. Edgar Dome by horizontal extension (after Kloppenburg et al., 2001, with permission from Elsevier). (a) Pre 3.33 Ga gabbro/diorite and dolerite intrusions, NE–SW extension, and doming of the Mt. Edgar Granitoid Complex. (b) Differential extension on the Mt. Edgar Shear Zone (MESZ) and lateral motion along a transfer zone within the granitoid complex at 3.31 Ga. (c) Final localized normal displacements and steepening of the MESZ followed by discordant intrusions of post-extensional plutons. Kearey et. Al., 2008. Ch.11 Precambrian Tectonics, p. 360

created by dense greenstones overlying buoyant, partially molten basement triggers the rise of granitoid domes at 3.31 Ga by solid state flow during extension (Fig. 14b). This extension is accommodated by displacement on the Mt. Edgar shear zone and by lateral strike-slip motion in a transfer zone within the gneissic basement. Normal-sense displacements drop greenstones down between the rising domes. Emplacement of the domes steepens the detachments and changes the geometry of the system so that its structure no longer resembles that of Phanerozoic core complexes (Fig. 14c). Steepening during periods of shortening provides an alternative explanation of the near vertical sides of the granitoid domes.

Both vertical and horizontal models involve several uncertainties. Problems with diapiric models commonly include uncertainties surrounding the timing of convective overturn and how an inverted density profile could be maintained or periodically established over a 750 million year history without thrust faulting (Van Kranendonk et al., 2004). How the stiff rheology of granitoids allows diapirism also is unclear. Problems with horizontal tectonic models may include a lack of evidence of large-scale tectonic duplication of the greenstones by thrusts in some areas and uncertainties surrounding how the formation of metamorphic core complexes could produce the distinctive ovoid patterns of the granitoids. Horizontal tectonic models also commonly encounter difficulty explaining the kinematics and horseshoe-shaped geometry of shear zones that border many granitoid domes (Marshak, 1999).

Archean – Proterozoic Boundary

During Late Archean and Early Proterozoic there is a distinct infer of plate tectonic pattern, from vertical to horizontal. The Archean thick continental proto-crust was uplifted, eroded and filled up with abundant clastic debris. Those series where further intruded by magmatic plumes and dykes. The most important driving force for tectonic pattern change is the secular cooling of the Earth. Predicated on the declining heat flux, the preferred model for the tectonic evolution of the Earth is therefore as follows (Hargraves, 1978) (Fig. 15). Such a tectonic evolution model is appraised on the basis of geologic evidence.

a) Following accretion and some degree of outward differentiation of sialic material the earth was rapidly brought to a state in which proto-crust was continuously being produced and consumed above vigorously circulating small-scale convection cells. Although positively buoyant, the crust was pulled down by the viscous drag forces associated with the down currents. (Fig. 15(1))

b) The viscous drag subduction slowly led to the accumulation of a globe-encircling, lithosphere scum-layer of intermediate composition (Fig. 15 (2)) and moderate thickness (10-20 km), completely decoupled from a still vigorously convecting asthenosphere-mantle below (Fig. 15 (3)). Its thickness constrained by the geothermal gradient, eruption of basalt above upwelling zones is no longer accompanied by lateral spreading as buoyant proto-crust is continuous. Outpouring of basalt onto the crust causes thickening from above is accompanied by remelting and removement of the crust at the base (Fig. 15 (4)).

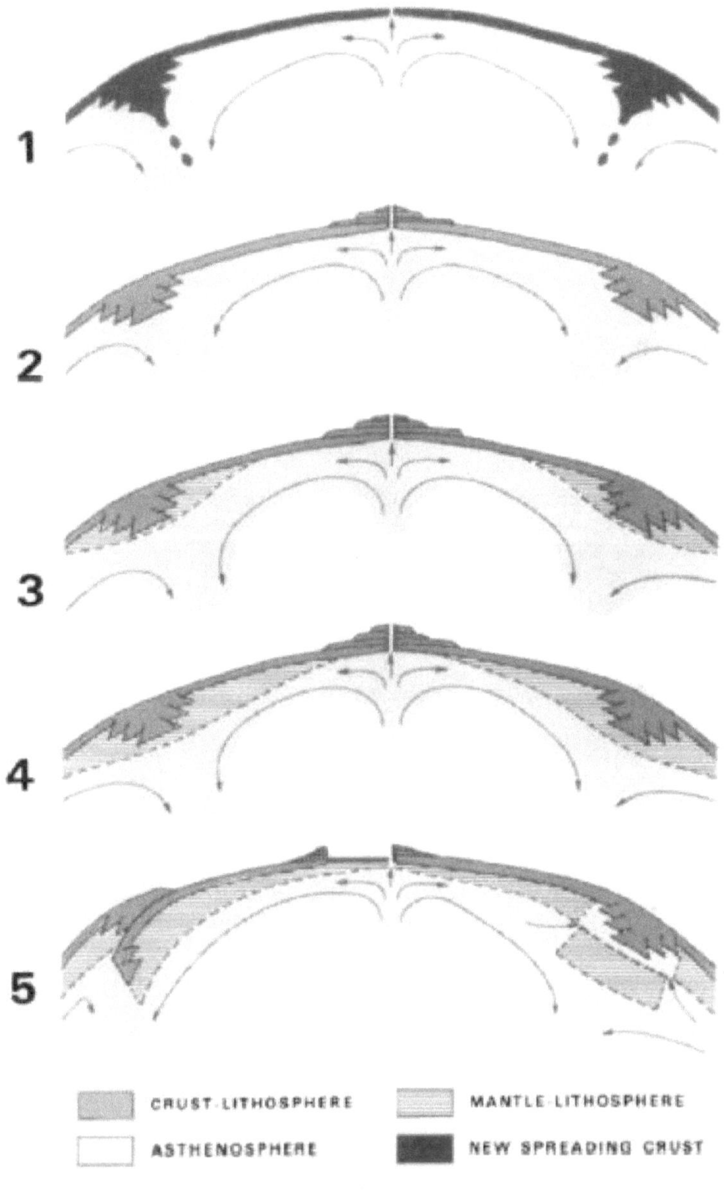

Fig. 15 Pictorial cartoon illustrating postulated evolution of tectonic style. R. B. Hargraves. 1981. PRECAMBRIAN PLATE TECTONICS. Ch. 2 Precambrian tectonic style: a liberal uniformitarian interpretation, p.46

c) With further decline in heat flow and increase in the mantle fraction of the lithosphere two distinct tectonic styles evolved : (1) The combination of thin, positively buoyant crust and thick, negatively buoyant mantle lithosphere necessary to give net negative buoyancy and permit subduction eventually arose somewhere on the globe. The buoyancy-powered plate-tectonic regime was initiated. (Fig. 15 (5) left) (2) Elsewhere, foundering of mantle-lithosphere slabs decoupled from thicker crust above also occurred and caused intracontinental orogeny. (Fig. 15 (5) right)

d) Constrained by the ambient thermal gradients, continental crust thickened by orogeny and sialic underplating simultaneous with the enlargement of the oceans.

On the other hand latest studies for the Archean orogeny tectonics support marginal orogeny pattern due to plate subduction instead of older models intra-plate orogeny. There are no sufficient evident to exclude one of those models.

Proterozoic Eon 2,5 – 0,6 Ga

Proterozoic Eon is the last and longer period of Precambrian. Rigid plates are currently formed and stable Cratons are then produced. On the basis of available data, Condie (2005b) concluded that 39% of the continental crust formed in the Archean, 31% in the Early Proterozoic, 12% in the Middle–Late Proterozoic, and 18% in the Phanerozoic. Some greenstone belts from the Archean period continue to form. A comparison of igneous rocks in Archean and Proterozoic belts indicates a progressive change in the bulk composition of the crust through time (Condie, 2005b). The large scale plate production and destruction rates are now minimized. Crust is decoupled from deeper convecting mantle. There are clear evince that orogeny cycles follow modern plate tectonics, Wilson Cycle. Continental break up, dispersal and collision is main feature of this period.

General Geology of Proterozoic Cratons

Two groups of metamorphic rocks consist the Proterozoic regions which are distinguished on the basis of their metamorphic grade and deformation history.

The first group consists of thick sequences of weakly deformed, un-metamorphosed sedimentary and volcanic rocks that were deposited in large basins on top of Archean cratons. The most common lithological assemblage in the weakly deformed parts of Early–Middle Proterozoic crust are quartzite-carbonate-shale sequences that reach thicknesses of some 10 km (Condie, 1982b). Other rock types that are either rare or absent in Archean belts appeared at this time, including extensive evaporites, phosphorous-rich sedimentary sequences, and red bed deposits. These latter rocks generally are interpreted to have accumulated in stable, shallow water environments after 2.0 Ga. The appearance and the preservation of such thick sequences of sedimentary rock has been interpreted to reflect the stabilization of Precambrian continental crust during Proterozoic times (Eriksson et al., 2001, 2005). By 1.8 Ga, the existence of large, stable landmasses and free oxygen in the Earth's atmosphere allowed all of the well-known sedimentary environments that characterize Phanerozoic times to develop (Eriksson et al., 2005).

The second group is composed of highly deformed, high-grade metamorphic rocks that define large orogenic belts. The highly deformed regions of Proterozoic crust are divisible into two types (Kusky & Vearncombe,1997). The first type consists of thick sedimentary sequences that were deformed into linear fold-and-thrust belts similar to those in Phanerozoic orogens. The second type consists of high-grade gneisses of the granulite and upper amphibolite facies. Some of the largest and best known of these latter belts form the ~1.0 Ga Grenville provinces of North America, South America, Africa, Antarctica, India, and Australia. Some contain ophiolites that resemble Phanerozoic examples except for the lack of highly deformed mantle-derived rocks at their bases in ophiolites older than ~1 Ga (Moores, 2002). The presence of these features reflects the importance of subduction, collision, and terrain accretion along Proterozoic continental margins (Carr et al., 2000; Karlstrom et al., 2002).

Both these groups contain distinctive suites of igneous rocks.

Proterozoic Plate tectonics

As observed through isotopic age determination studies it is suggested that crust production was episodic with rapid net growth at 2.7, 1.9, and 1.2 Ga and slower growth afterward (Condie, 2000; Rino et al., 2004). Tectonic stability refers to the general resistance of the Cratons to large-scale lithospheric recycling processes. Two of the most important mechanisms of Late Archean and Early Proterozoic continental growth and cratonic root evolution were magma addition and terrain accretion.

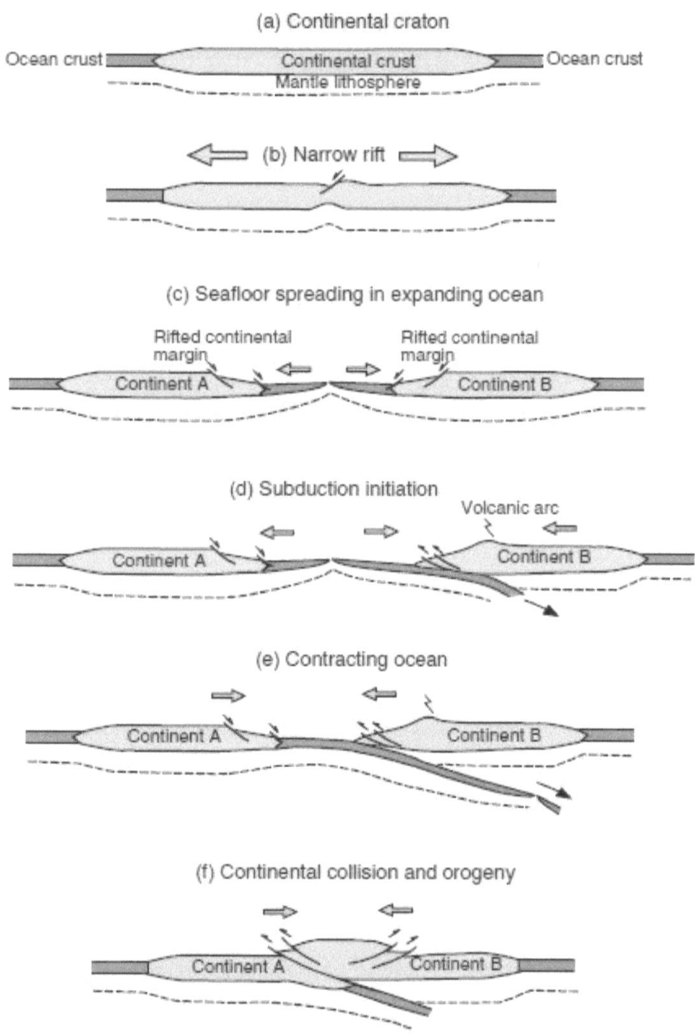

Fig. 16 The Wilson cycle: (a) continental craton; (b) formation of a narrow rift; (c) initiation of seafloor spreading and formation of rifted continental margins in an expanding ocean basin; (d) initiation of subduction; (e) closing of ocean basin; (f) continental collision and orogeny. Kearey et. Al..2008. Global Tectonics. Chapter 7, p.208

The appearance of the metamorphic products of subduction and continental collision during the Early Proterozoic, including eclogites and other >1 GPa high pressure metamorphic assemblages, represents an important marker of the onset of tectonic processes similar to those seen in the Phanerozoic Earth (O'Brien & Rötzler, 2003; Collins et al., 2004; Brown, 2006). Moreover the presence of ophiolitic assemblages in Precambrian orogens provides another possible marker of tectonic processes similar to those operating during the Phanerozoic-Wilson Cycle (Fig. 16) (de Wit, 2004; Parman et al., 2004). The oldest examples are Early Proterozoic in age and support interpretations that seafloor spreading and associated ocean crust formation was an established mechanism of plate tectonics by this time (Kearey et. Al., 2008).

Supercontinental Cycle

The several stable Cratons produced during Precambrian coalesced into large landmasses called Supercontinents based on geologic evidence for the repeated occurrence of continental rifting and collision since the Late Archean.

Late Archean – Early Proterozoic Supercontinents

Recording the earliest Supercontinents, about 35 Archean Cratons are observed and most of them display rifted margins, indicating that they fragmented from larger landmasses (Bleeker, 2003). Several possible scenarios have been envisioned for the global distribution of the cratons during the transition from Late Archean to Early Proterozoic times. These possibilities include a single supercontinent, called Kenorland by Williams et al. (1991) after an orogenic event in the Canadian Superior Province, and the presence of either a few or many independent aggregations called Supercratons (Fig. 17 (a)). On the other hand Bleeker (2003) concluded that the degree of geologic similarity among the exposed cratons favors the presence of several transient (Fig. 17 (b)), more or less independent Supercratons rather than a single supercontinent or even a many small dispersed landmasses model, which is rather unlikely (Fig. 17 (c)).

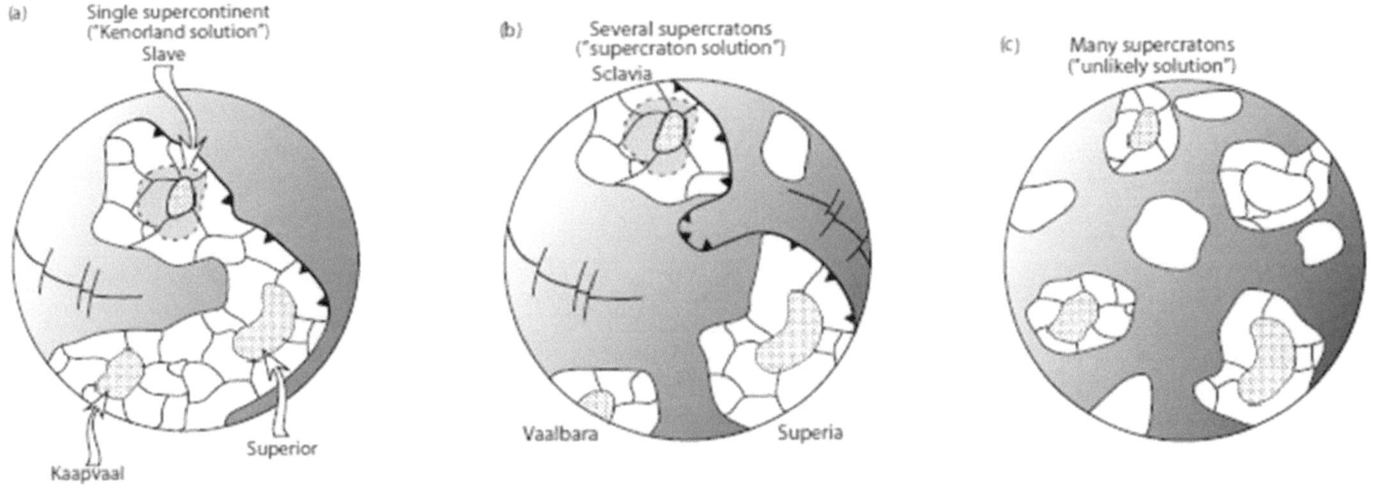

Fig. 17 Cartoons representing possible craton configurations during Late Archean–Early Proterozoic times. (a) William's Model: One Single Supercontinent Kenorland; (b) Bleeker's Model Three well-known cratons, Slave, Superior and Kaapvaal; (c) Bleeker's many small dispersed landmasses Model (after Bleeker, 2003, with permission from Elsevier). Kearey et. Al..2008. Global Tectonics. Ch. 11, p.374

Late Proterozoic Supercontinents

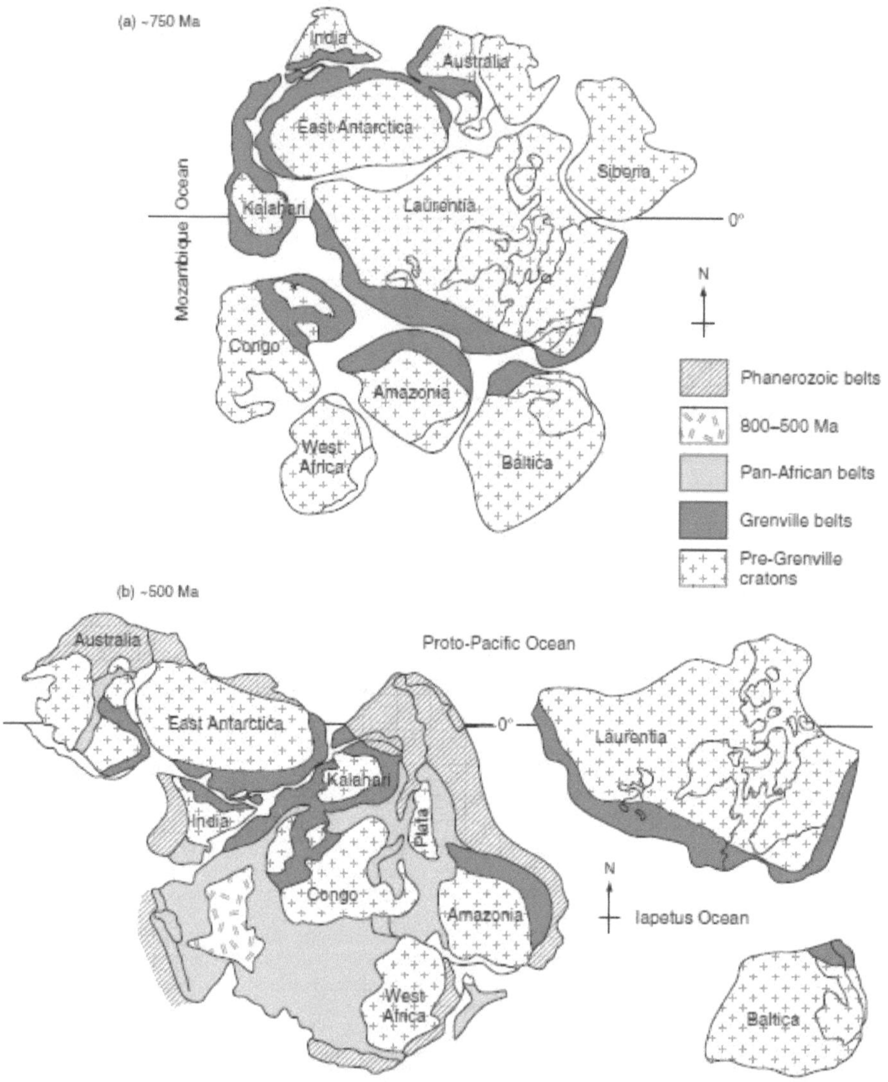

Fig. 18 (a) Reconstruction of the Late Proterozoic supercontinent Rodinia. (b) Late Cambrian paleogeography after the break-up of Rodinia and the formation of Gondwana (after Hoffman, 1991, with permission from Science 252, 1409–12 with permission from the AAAS). Kearey et. Al..2008. Global Tectonics. Ch. 11. p.371

The age of the oldest sedimentary rock associated with breakup, and the provinciality of certain animal groups across the split, suggest that the supercontinent fragmented at about 750 Ma (Storey, 1993). During fragmentation the blocks now making up East Gondwana (East Antarctica, Australia, and India) moved anticlockwise, opening the proto-Pacific Ocean (Panthalassa), and collided with the blocks of West Gondwana (Congo, West Africa, and Amazonia). The intervening Mozambique Ocean closed by the pincer-like movements of these blocks and Gondwana was created when they collided to form the Mozambique belt of East Africa and Madagascar. Gondwana then rotated clockwise away from Laurentia about 200 Ma later. Southern Africa was located at the pivot of these movements and Baltica moved independently away from Laurentia, opening the Iapetus Ocean, which subsequently closed during the assembly of Pangea.

This model is mostly based on geological evidence. Paleomagnetic tests add new constraints to this hypothesis. There are now considerable paleomagnetic evidence that the cratons of Africa and South America (Fig. 18), except Amazonia, were never part of Rodinia, although they were probably close to it. The first paleomagnetic test of the SWEAT (South West US and East AnTarctica) hypothesis was carried out by Powell et al. (1993) who showed that paleomagnetic poles at 1055 Ma and at 725 Ma for Laurentia and East Gondwana are in agreement when repositioned according to the Rodinia reconstruction, thereby lending support to the hypothesis. Between 725 Ma and the Cambrian the APWPs diverge, suggesting that East Gondwana broke away from Laurentia after 725 Ma. The only fragment of Rodinia for which a detailed Apparent Polar Wanderer (APW) path can be defined for the period 1100–725 Ma is Laurentia (McElhinny & McFadden, 2000).

Moreover, West Gondwana formed when many small ocean basins that surrounded the African and South American cratons closed during the opening of the proto-Pacific Ocean, creating the Pan-African orogens (Fig. 18b). Subsequent closure of the Mozambique Ocean resulted in the collision and amalgamation of West Gondwana with the blocks of East Gondwana. This amalgamation may have created a short-lived Early Cambrian supercontinent called Pannotia. The existence of this supercontinent is dependent on the time of rifting between Laurentia and Gondwana (Cawood et al., 2001). Models of Pannotia (Fig. 19) are based mostly on geologic evidence that Laurentia and Gondwana were attached or in close proximity at the end of the Late Proterozoic (Dalziel, 1997). However, the paleomagnetic poles for these two landmasses do not overlap, suggesting that an alternative configuration where Laurentia is separated from Gondwana at this time also is possible (Meert & Torsvik, 2003).

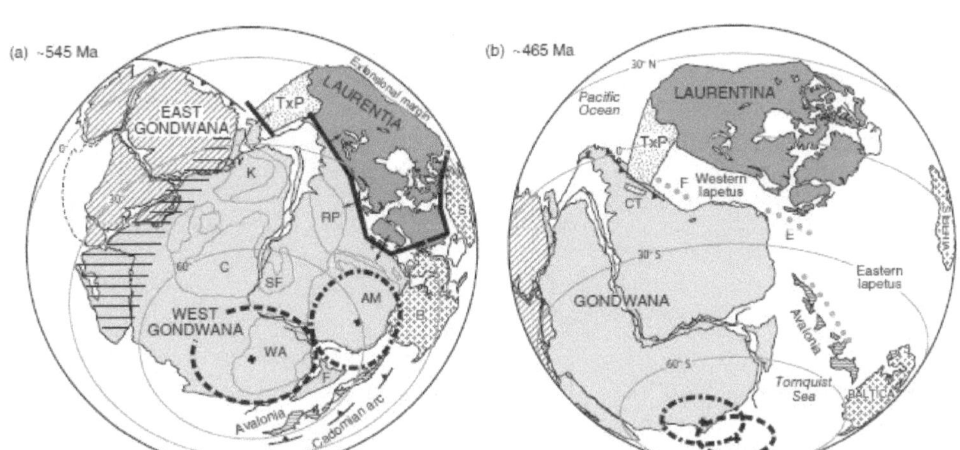

Fig. 19 Postulated reconstructions of (a) Pannotia at ~545 Ma and (b) the rifting of Laurentia and Gondwana at ~465 Ma emphasizing the paleogeography of Laurentia relative to Gondwana (after Dalziel, 1997, with permission from the Geological Society of America) Kearey et. Al..2008. Global Tectonics. Ch. 11, p.3375

Another controversial aspect of the Rodinia supercontinent concerns the effect of its dispersal on past climates. Some studies suggest that as Rodinia fragmented the planet entered an icehouse or snowball Earth state in which it was intermittently completely covered by ice (Evans, 2000; Hoffman & Schrag, 2002). The geologic evidence for this intermittent but widespread glaciation includes glacial deposits of Late Proterozoic age that either contain carbonate debris or are directly overlain by carbonate rocks indicative of warm waters. In addition, paleomagnetic data suggest that during at least two Late Proterozoic glacial episodes ice sheets reached the equator. One possible explanation of these observations is that periods of global glaciation during the Late Proterozoic were controlled by anomalously low atmospheric CO_2 concentrations (Hyde et al., 2000; Donnadieu et al., 2004). During break-up, the changing paleogeography of the continents may have led to an increase in runoff, and hence consumption of CO_2, through continental weathering that decreased atmospheric CO_2 concentrations. The extreme glacial conditions may have ended when volcanic outgassing of CO_2 produced a sufficiently large greenhouse effect to melt the ice. The resulting "hothouse" would have enhanced precipitation and weathering, giving rise to the deposition of carbonates on top of the glacial deposits during sea-level (Hoffman et al., 1998). Alternatively, these transitions may have resulted mainly from the changing configuration of continental fragments and its effect on oceanic circulation. Whichever view is correct, these interpretations suggest that the break-up of Rodinia triggered large changes in global climate (Kearey et. Al., 2008). However, the origin, extent, and termination of the Late Proterozoic glaciations, and their possible relationship to the supercontinental breakup, remains an unresolved and highly contentious issue (Kennedy et al., 2001; Poulsen et al., 2001).

SUMMARY

Precambrian Era explores the greatest overall Earth's changes, either in composition, boundary conditions, geology or tectonics. There have been several studies to determine exact conditions and many models have been formed. Nevertheless, it's not yet safely determined which of the models formed best describe the Precambrian conditions. However there are some general thoughts accepted about each period of Precambrian (Fig. 20). During Hadean Eon, no horizontal tectonics are marked, only distinctive primordial differentiation of magma. Moving on to Archean, small scale convection is taking place along with vigorous high rates of production and destruction of lithosphere. During Late Archean to Early Proterozoic Boundary tectonic patterns gradually change from vertical to horizontal due to terrestrial heat loss. While in the end of Precambrian, tectonics are approaching the modern ones. Wilson Cycle controls Supercontinental fragmentation, dispersal and collision.

Hadean Eon 4500-3900 Ga	Controversial - Formation of proto-crust and proto-lithosphere of non successfully determined composition. No horizontal tectonics.
Archean Eon 3900-2700 Ga	High rate of radiogenic decay, small-scale mantle convection and vigorous high formation-destruction rates of oceanic crust which was subducted. Extensive rifting in marginal basins.
Archean - Proterozoic Boundary 2700-2300 Ga	Transitional and diachronous period during which much uplift and erosion of late Archaean thick continental crust. Stabilization of crust-lithosphere following aggregation of Archaean proto-plates. Many dykes and stratiform intrusions injected into stabilizing crust.
Proterozoic Eon 2300-600 Ga	Lower mantle temperatures and slower global spreading rate leads to modern Wilson-cycle from early Proterozoic. Narrow geosynclines and mobile belts giving rise to extensive more stable plates. First Supercontinents.
Phanerozoic Eon 600 Ma to Present	Wilson cycle controls all geological-tectonic phenomena

Fig. 20 Alternated Table by Anastasia Kokkinou after Brian F. Widdley. A. Kroener. PRECAMBRIAN PLATE TECTONICS.1981. Chapter 1, p.16

The continuous questioning about the nature of tectonics through time, uniformitarian or non uniformitarian for Precambrian still lies. There have been a differentiation along the three periods of Precambrian to determine this issue. A non uniformitarian tectonic model is characterizing the Hadean Eon. For the

Archean Eon a moderate uniformitarian would better fit. Considering the much different current boundary conditions, it couldn't be possible follow the same tectonic models as in modern Earth. When in Proterozoic terrestrial heat flow is lower and global tectonic spreading is slower, a rather uniformitarian to modern tectonic model is suitable.

Bibliography

A. Kroener (Editor) et. Al.. 1981. Developments in Precambrian Geology 4 – PRECAMBRIAN PLATE TECTONICS. ELSEVIER SCIENTIFIC PUBLISHING COMPANY. Institut fuer Geowissenschaften, Johannes Gutenberg-Universitgt Mainz, Mainz, Federal Republic of Germany. Chapters 1, 3, 4, 18, 23 & 26.

Tracy Rushmer, John Adam, Jonathan O'Neil and Don Francis. 2013. Hadean greenstones from the Nuvvuagittuq fold belt and the origin of the Earth's early continental crust. © 2012 Geological Society of America, Geology 2012;40;363-366, doi: 10.1130/G32623.1. 1GEMOC, Earth and Planetary Sciences, Macquarie University, NSW 2109, Australia, Department of Terrestrial Magmatism, Carnegie Institution of Washington, 5241 Broad Branch Road N.W., Washington, D.C. 20015, USA, Earth and Planetary Sciences Department, McGill University, 3450 University Street, Montreal, Quebec H3A 2A7, Canada

Tracy Rushmer, Simon Turner, Mark Reagan, Jean-Francois Moyen. 2015(Not yet published). Hadean Subduction. Department of Earth and Planetary Sciences, Macquarie University, Sydney, Australia, Department of Geoscience, University of Iowa, Iowa City, Iowa 52242, USA UMR 6524 CNRS and Universite Jean-Monnet, 23 rue du Dr. Michelon, Saint-Etienne, France.

Philip Kearey, Keith A. Klepeis, Frederick J. Vine. 2008. Global Tectonics. John Wiley & Sons Ltd Publications UK. Chapter 11 – Precambrian Tectonics.

References Cited

Andreasen, R., and Sharma, M., 2009, Neodymium-142 evidence for Hadean mafic crust: Comment: Science, v. 325, p. 267–269, doi:10.1126/science .1169604.

Arndt, N.T., Lewin, E. & Albarède, F. (2002) Strange partners: formation and survival of continental crust and lithospheric mantle. In Fowler, C.M.R., Ebinger, C.J. & Hawkesworth, C.J. (eds) The Early Earth: physical, chemical and biological development. Spec. Pub. geol. Soc. Lond. 199, 91–103.

Bickle, M.J., Nisbet, E.G. & Martin, A. (1994) Archean greenstone belts are not oceanic crust. J. Geol. 102, 121–38.

Bleeker, W. (2003) The late Archean record: a puzzle in ca. 35 pieces. Lithos 71, 99–134.

Blewitt, R.S. (2002) Archaean tectonic processes: a case for horizontal shortening in the North Pilbara Granite–Greenstone Terrane, Western Australia. Precambrian Res. 113, 87–120.

Blichert-Toft, J., and Albarede, F., 2008, Hafnium isotopes in Jack Hills zircons and the formation of the Hadean crust: Earth and Planetary Science Letters, v. 265, p. 686–702, doi:10.1016/j.epsl.2007.10.054.

Bowring, S.A. & Williams, I.S. (1999) Priscoan (4.00–4.03 Ga) orthogneisses from northwestern Canada. Contrib. Mineral. Petrol. 134, 3–16.

Brown, M. (2006) Duality of thermal regimes is the distinctive characteristics of plate tectonics since the Neoarchean. Geology 34, 961–4.

Buick, R. (2001) Life in the Archaean. In Briggs, D.E.G. & Crowther, P.R. (eds) Palaeogeology II, pp. 13–21. Blackwell, Oxford.

Carlson, R.W., Pearson, D.G. &. James, D.E. (2005) Physical, chemical, and chronological characteristics of continental mantle. Rev. Geophys. 43, RG1001, doi:10.1029/2004RG000156.

Carr, S.D. et al. (2000) Geologic transect across the Grenville orogen of Ontario and New York. Can. J. Earth Sci. 37, 193– 216.

Cawood, P.A., Kröner, A. & Pisarevsky, S. (2006) Precambrian plate tectonics: criteria and evidence. GSA Today 16, 4–11.

Cawood, P.A., McCausland, P.J.A. & Dunning, G.R. (2001) Opening Iapetus: constraints from the Laurentian margin in Newfoundland. Bull. geol. Soc. Am. 113, 443–53.

Chardon, D., Choukroune, P. & Jayananda, M. (1996) Strain patterns, décollement and incipient sagducted greenstones terrains in the Archaean Dharwar craton (south India). J. struct. Geol. 18, 991–1004.

Collins, A.S. et al. (2004) Temporal constraints on Palaeoproterozoic eclogite formation and exhumation (Usagaran Orogen, Tanzania). Earth planet. Sci. Lett. 224, 175–92.

Collins, W.J., Van Kranendonk, M.J. & Teyssier, C. (1998) Partial convective overturn of Archaean crust in the east Pilbara Craton, Western Australia: driving mechanisms and tectonic implications. J. struct. Geol. 20, 1405–24.

Condie, K.C. (2005a) TTGs and adakites: are they both slab melts? Lithos 80, 33–44.

Condie, K.C. (2005b) Earth as an Evolving Planetary System. Elsevier, Amsterdam.

Condie, K.C. (2000) Episodic continental growth models: afterthoughts and extensions. Tectonophysics 322, 153–62.

Condie, K.C. (1982b) Early and middle Proterozoic supracrustal successions and their tectonic settings. Amer. J. Sci. 282, 341– 57.

Dalziel, I.W.D. (1997) Neoproterozoic–Paleozoic geography and tectonics: review, hypothesis, environmental speculation. Bull. geol. Soc. Am. 109, 16–42.

David Jean, Laurent Godin, Ross Stevenson, Jonathan O'Neil und Don Francis: U-Pb ages (3.8–2.7 Ga) and Nd isotope data from the newly identified Eoarchean Nuvvuagittuq supracrustal belt, Superior Craton, Canada. GSA Bulletin, Bd. 121; Nr. 1-2; S. 150-163; January 2009, doi:10.1130/B26369.1

de Wit, M.J. (2004) Archean Greenstone Belts do contain fragments of ophiolites. In Kusky, T.M. (ed.) Precambrian Ophiolites and Related Rocks. Developments in Precambrian Geology, 13, pp. 599–613. Elsevier, Amsterdam.

Dhuime, B., Hawkesworth, C.J., Cawood, P.A., and Storey, C.D., 2012, A change in the geodynamics of continental growth 3 billion years ago: Science, v. 335, p. 1334–1336, doi:10.1126/science.1216066.

Donnadieu, Y. et al. (2004) A "snowball Earth" climate triggered by continental break-up through changes in runoff. Nature 428, 303–6.

Drummond, M.S., and Defant, M.J., 1990, A model for trondhjemite-tonalitedacite genesis and crustal growth via slab melting: Archean to modern comparisons: Journal of Geophysical Research, v. 95, p. 21503–21521, doi:10.1029/JB095iB13p21503.

Ernst, R.E., Buchan, K.L. & Campbell, I.H. (2005) Frontiers in Large Igneous Province research. Lithos 79, 271–97.

Eriksson, P.G. et al. (2005) Patterns of sedimentation in the Precambrian. Sed. Geol. 176, 17–42

Erickson, S.G., Strayer, L.M. & Suppe, J. (2001) Initiation and reactivation of faults during movements over a thrust-fault ramp: numerical mechanical models. J. struct. Geol. 23, 11–23.

Eriksson, P.G. et al. (2001) An introduction to Precambrian basins: their characteristics and genesis. Sed. Geol. 141–142, 1–35.

Evans, D. (2000) Stratigraphic, geochronological, and paleomagnetic constraints upon the Neoproterozoic climatic paradox. Amer. J. Sci. 300, 347–433.

Guitreau, M., Blitchert-Toft, J., Mojzsis, S.J., Roth, A.S.G., and Bourdon, B., 2013, A legacy of Hadean silicate differentiation inferred from Hf isotopes in Eoarchean rocks of the Nuvvuagittuq supracrustal belt (Quebec, Canada): Earth and Planetary Science Letters, v. 362, p. 171–181, doi:10.1016/j.epsl.2012.11.055.

Hargraves, R.B. (1986) Faster spreading or greater ridge length in the Archean? Geology 14, 750–2.

Harrison, M.T., 2009, The Hadean crust: Evidence from >4 Ga zircons: Annual Review of Earth and Planetary Sciences, v. 37, p. 479–505, doi: 10.1146/annurev.earth.031208.100151.

Herzberg, C. (1999) Formation of cratonic mantle as plume residues and cumulates. In Fei, Y., Bertka, C. & Mysen, B.O. (eds) Mantle Petrology: fi eld observations and high pressure experimentation, pp. 241–57, Geochemical Society, Houston, TX.

Hoffman, P.F. & Schrag, D.P. (2002) The snowball Earth hypothesis: testing the limits of global change. Terra Nova 14, 129–55.

Hoffman, P.F. et al. (1998) A Neoproterozoic snowball Earth. Science 281, 1342–6.

Hyde, W.T. et al. (2000) Neoproterozoic "snowball Earth" simulations with a coupled climate/ice-sheet model. Nature 405, 425–9.

Ishizuka, O., Tani, K., Reagan, M.K., Kanayama, K., Umino, S., Harigane, Y., Sakamoto, I., Miyajima, Y., Yuasa, M., and Dunkley, D.J., 2011, The time scales of subduction initiation and subsequent evolution of an oceanic island arc: Earth and Planetary Science Letters, v. 306, p. 229–240, doi:10.1016/j.epsl.2011.04.006.

Ishizuka, O., 2008, Volcanic and tectonic framework of the hydrothermal activity of the Izu-Bonin arc: Resource Geology, v. 58, p. 206–219.

Karlstrom, K.E. et al. (2002) Structure and evolution of the lithosphere beneath the Rocky Mountains: initial results from the CD-ROM Experiment. GSA Today 12, 4–10.

Kemp, A.I.S., Wilde, S.A., Hawkesworth, C.J., Coath,C.D., Nemchin, A., Pidgeon, R.T., Vervoot, J.D., and DuFrane, S.A., 2010, Hadean crustal evolution revisited: New constraints from Pb-Hf isotope systematics of the Jack Hills zircons: Earth and Planetary Science Letters, v. 296, p. 45–56, doi:10.1016/j.epsl.2010.04.043.

Kennedy, M.J., Christie-Blick, N. & Prave, A.R. (2001) Carbon isotopic composition of Neoproterozoic glacial carbonates as a test of paleoceanographic models for snowball Earth phenomena. Geology 29, 1135–8.

King, S.D. (2005) Archean cratons and mantle dynamics. Earth planet. Sci. Lett. 234, 1–14.

Kloppenburg, A., White, S.H. & Zegers, T.E. (2001) Structural evolution of the Warrawoona Greenstone Belt and adjoining granitoid complexes, Pilbara craton, Australia: implications for Archaean tectonic processes. Precambrian Res. 112, 107–47.

Kröner, A. (1981) Precambrian plate tectonics. In Kröner, A. (ed.) Precambrian Plate Tectonics, pp. 57–90. Elsevier, Amsterdam.

Kusky, T.M. & Vearncombe, J.R. (1997) Structural Aspects. In de Wit, M.J. & Ashwal, L.D. (eds) Greenstone Belts, pp. 91–124. Clarendon Press, Oxford, UK.

Lenardic, A. (1998) On the partitioning of mantle heat loss below oceans and continents over time and its relationship to the Archaean paradox. Geophys. J. Int. 134, 706–20.

Mareschal, J.-C. & Jaupart, C. (2006) Archean thermal regime and stabilization of the cratons. In Benn, K., Mareschal, J.C., & Condie, K.C. (eds) Archean Geodynamics and Evironments. Geophys. Monogr. Ser. 164, pp. 61–73. American Geophysical Union, Washington, DC.

Marshak, S. (1999) Deformation style way back when: thoughts on the contrasts between Archean/Paleoproterozoic and contemporary orogens. J. struct.Geol. 21, 1175–82.

Martin, H. (1986) Effects of steeper Archean geothermal gradient on geochemistry and subduction-zone magmas. Geology 14, 753–6.

McElhinny, M.W. & McFadden, P.L. (2000) Paleomagnetism: continents and oceans. Academic Press, San Diego.

Meert, J.G. & Torsvik, T.H. (2003) The making and unmaking of a supercontinent: Rodinia revisited. Tectonophysics 375, 261– 88.

Mojzsis, S.J., Harrison, T.M., and Pigeon, R.T., 2001, Oxygen-isotope evidence from ancient zircons for liquid water at the Earth's surface 4300 Myr ago: Nature, v. 409, p. 178–181, doi: 10.1038/35051557.

Moores, E.M. (2002) Pre–1 Ga (pre-Rodinian) ophiolites: their tectonic and environmental implications. Bull. geol. Soc. Am.114, 80–95.

Moyen, J.-F., 2013, Archaean tectonic systems: A naive geochemist's view: Vienna, European Geosciences Union General Assembly 2013, abs. EGU2013-2331.

Nisbet, E.G. et al. (1993) Constraining the potential temperature of the Archean mantle: a review of the evidence from komatiites. Lithos 30, 291–307.

O'Brien, P.J. & Rötzler, J. (2003) High pressure granulites: formation, recovery of peak conditions and implications for tectonics. J. metam. geol. 21, 3–20.

O'Neil, J., Francis, D., and Carlson, R.W., 2011, Implications of the Nuvvuagittuq Greenstone Belt for the formation of Earth's early crust: Journal of Petrology, v. 52, p. 985–1009, doi: 10.1093/petrology/egr014. Foley, S.F., Buhre, S. & Jacob, D.E. (2003) Evolution of the Archaean crust by delamination and shallow subduction. Nature 421, 249–52

O'Neil, J., Carlson, R.W., Francis, D., and Stevenson, R.K., 2008, Neodymium- 142 evidence for Hadean mafi c crust: Science, v. 321, p. 1828–1831, doi:10.1126/science.1161925.

O'Reilly, S.Y. et al. (2001) Are lithospheres forever? Tracking changes in subcontinental lithospheric mantle through time. GSA Today 11, 4–10.

Parman, S.W., Grove, T.L. & Dann, J.C. (2004) A subduction origin for komatiites and cratonic lithospheric mantle. S. Afr. J. Geol. 107, 107–18.

Pearce, J.A., 2008, Geochemical fi ngerprinting of oceanic basalts with application to ophiolite classifi cation and the search for Archean oceanic crust: Lithos, v. 100, p. 14–48, doi:10.1016/j .lithos.2007.06.016.

Pearson, D.G. et al. (2002) The development of lithosperic keels beneath the earliest continents: time constraints using PGE and Re–Os isotope systematics. In Fowler, C.M.R., Ebinger, C.J. &

Hawkesworth, C.J. (eds) The Early Earth: physical, chemical and biological development. Spec. Pub. geol. Soc. Lond. 199, 65–90.

Poulsen, C.J. et al. (2001) Response of the mid-Cretaceous global ocean circulation to tectonic and CO2 forcings. Paleoceanography, 16, 576–92.

Rapp, R.P., Shimizu, N., and Norman, M.D., 2003, Growth of early continental crust by partial melting of eclogite: Nature, v. 425, p. 605–609, doi:10.1038/ nature02031.

Reagan, M.K., and 15 others, 2010, Fore-arc basalts and subduction initiation in the Izu-Bonin-Mariana system: Geochemistry, Geophysics, Geosystems, v. 11, Q03X12, doi:10.1029/2009GC002871.

Rino, S. et al. (2004) Major episodic increase of continental crustal growth determined from zircon ages of river sands; implications for mantle overturns in the Early Precambrian. Phys. Earth planet. Interiors 146, 369–94.

Roth, A.S.G., Bourdon, B., Mojzsis, S.J., Touboul, M., Sprung, P., Guitreau, M., and Blitchert-Toft, J., 2013, Inherited 142Nd anomalies in Eoarchean protoliths: Earth and Planetary Science Letters, v. 361, p. 50–57, doi:10.1016/j.epsl.2012.11.023.

Rudnick, R.L. & Fountain, D.M. (1995) Nature and composition of the continental crust: a lower crustal perspective. Rev. Geophys. 33, 267–309.

Sleep, N.H. (2003) Survival of Archean cratonal lithosphere. J. geophys. Res. 108, 2302, doi:10.1029/2001JB000169.

Smithies, R.H., Champion, D.C., and van Kranendonk, M.J., 2009, Formation of Paleoarchean continental crust through infracrustal melting of enriched basalt: Earth and Planetary Science Letters, v. 281, p. 298–306, doi:10.1016/j .epsl.2009.03.003.

Stern, R.J., Reagan, M., Ishizuka, O., Ohara, Y., and Whattam, S., 2012, To understand subduction initiation, study forearc crust; to understand forearc crust, study ophiolites: Lithosphere, v. 4, p. 469–483, doi:10.1130/L183.1.

Stern, R.J., 2005, Evidence from ophiolites, blueschists, and ultrahigh pressure metamorphic terranes that the modern episode of subduction tectonics began in Neoproterozoic time: Geology, v. 33, p. 557–560, doi:10.1130/G21365.1.

Storey, B.C. (1993) The changing face of late Precambrian and early Palaeozoic reconstructions. J. geol. Soc. Lond. 150, 665–8.

Tessalina, S.G., Bourdon, B., Van Kranendonk, M., Birk, J., and Philippot, P., 2010, Infl uence of Hadean crust evident in basalts and cherts from the Pilbara Craton: Nature Geoscience, v. 3, p. 214–217, doi:10.1038/NGEO772.

Tomlinson, K.Y. & Condie, K.C. (2001) Archean mantle plumes: evidence from greenstone belt geochemistry. In Ernst, R.E. & Buchan, K.L. (eds) Mantle Plumes: their identifi cation through time. Geol. Soc. Amer. Sp. Paper 352, 341–57.

Van Kranendonk, M.J. et al. (2007) Secular tectonic evolution of Archean continental crust: interplay between horizontal and vertical processes in the formation of the Pilbara Craton, Australia. Terra Nova 19, 1–38.

Van Kranendonk, M.J. et al. (2004) Critical tests of vertical vs. horizontal tectonic models for the Archean East Pilbara Granite–Greenstone Terrane, Pilbara Craton, Western Australia. Precambrian Res. 131, 173–211.

Van Kranendonk, M.J. et al. (2002) Geology and tectonic evolution of the Archean North Pilbara Terrain, Pilbara Craton, Western Australia. Econ. Geol. 97, 695–732.

Windley, B.F. (1981) Precambrian rocks in the light of theplate-tectonic concept. In Kröner, A. (ed.) Precambrian PlateTectonics, pp. 1–20. Elsevier, Amsterdam.

Wilde, S.A. et al. (2001) Evidence from detrital zircons for the existence of continental crust and oceans on the Earth 4.4 Gyr ago. Nature 409, 175–8.

Williams, H. et al. (1991) Anatomy of North America: thematic geologic portrayals of the continent. Tectonophysics 187, 117–34.

Zegers, T.E. & van Keken, P.E. (2001) Middle Archean continent formation by crustal delamination. Geology 29, 1083–6.

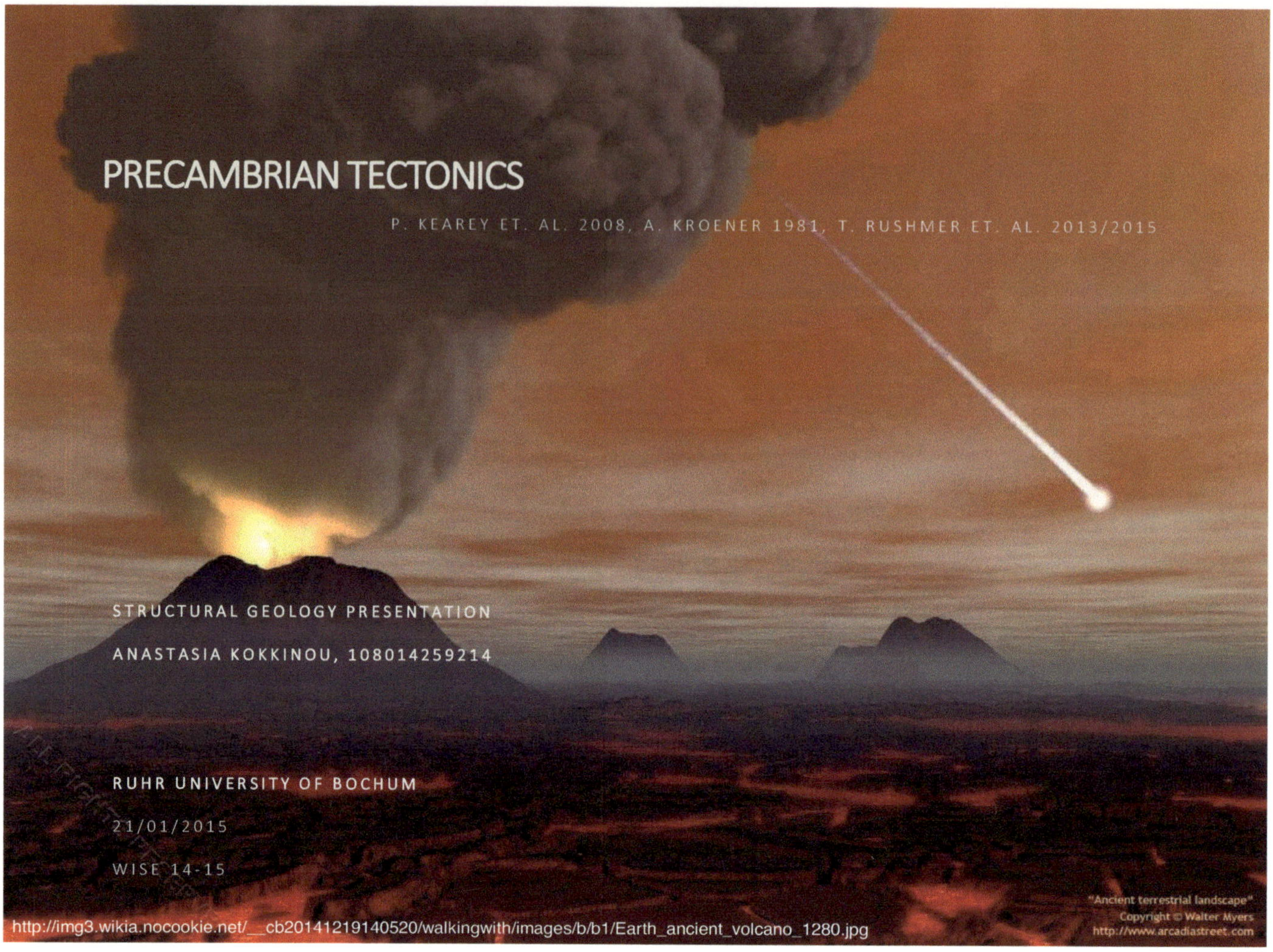

PRECAMBRIAN TECTONICS

P. KEAREY ET. AL. 2008, A. KROENER 1981, T. RUSHMER ET. AL. 2013/2015

STRUCTURAL GEOLOGY PRESENTATION

ANASTASIA KOKKINOU, 108014259214

RUHR UNIVERSITY OF BOCHUM

21/01/2015

WISE 14-15

http://img3.wikia.nocookie.net/__cb20141219140520/walkingwith/images/b/b1/Earth_ancient_volcano_1280.jpg

Introduction

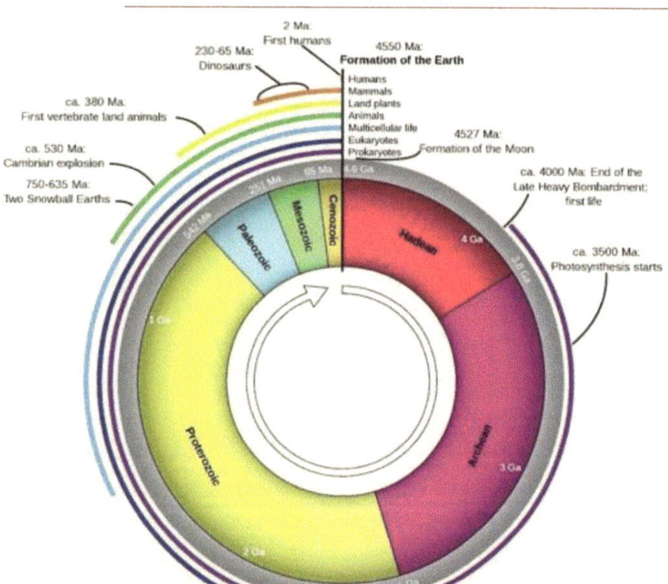

Fig. 1
https://bellavitabyjamie.wordpress.com/2012/07/21/timeline/earth-timeline/

Precambrian duration 4,5 Ga – 600 Ma, 7/8 Earth's History

Main Partition:

1. Hadean Eon 4,5 – 3,9 Ga
2. Archean Eon 3,9 – 2,5 Ga
3. Proterozoic Eon 2,5 Ga – 600 Ma

Main Events:

- Solid Earth Condition towards Stable Cratons Formation
- Oxygen-Rich Atmosphere Formation, First Form of Life, Water
- Primary Earth Tectonics

Uniformitarian or Non Uniformitarian?

Hadean Eon (4,5 – 3,9 Ga)

Controversial Theories:

Magma Ocean Earth	Crust
Giant Impact Theory	Zircon age data (4,4 Ga. *Wilde et al., 2001*)
Magma Ocean	TTGs (3,8 Ga. *Rushmer et. al., 2015*)

- Zircon Isotope Age – Problematic Origin Explanation due to co-existence with arc-like, mafic material.

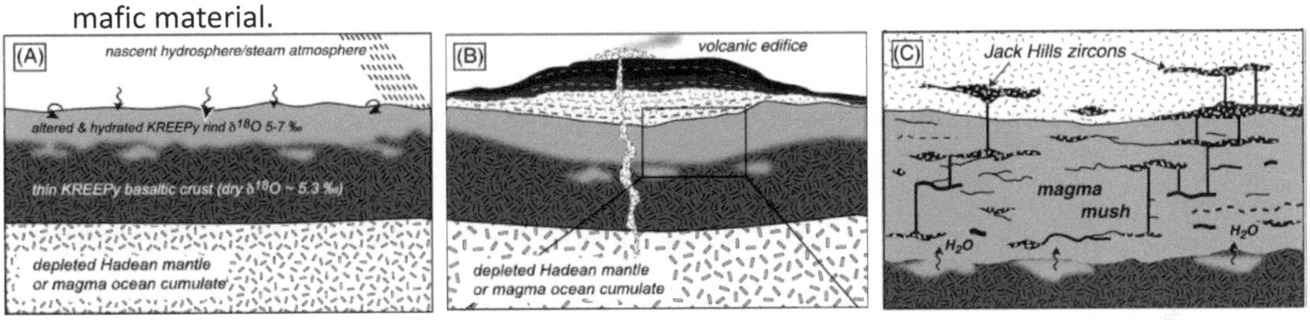

Fig. 2 Hadean Zircon initiation: From Jack Hills zircon study (Kemp et al., 2010), solidification and crystallization of a magma ocean (KREEPy layer) and a prolonged period of partial melting.

Oldest Crust:

TTGs (tonalite-trondhjemite-granodiorite) ~3,6Ga, Nuvvuagittuq Greenstones, Canada *(O'Neil et al, 2012, Roth et al., 2013, Guitreau et al., 2013).*

- Definite Hadean (primoridal) magma differentiation. Non horizontal tectonics (plate interaction) until Archean *(Kerrich & Polat, 2006; Shirey et al.2008; Polat, 2012).*
- Subduction initiation signatures. Maybe some type of plates and very limited horizontal tectonic implications *(Turner et. Al., 2015?).*

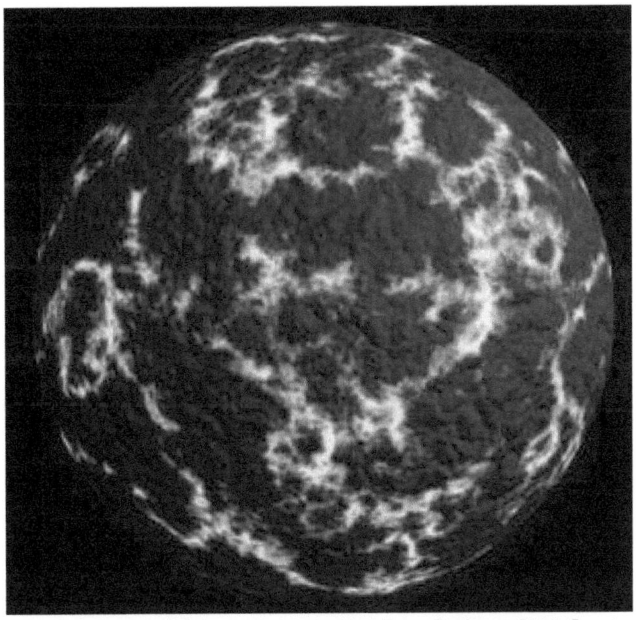

Fig. 3 Hadean Earth's image reconstruction. Tracy Rushmer. 2014. Onset of subduction and the genesis of TTGs (tonalities-trondhjemites-granodiorites) in the in the earliest Earth. (Powerpoint presentation)

Archean Eon (3,9-2,5 Ga)

- Vigorous growth (accretion) and destruction of oceanic lithosphere
- Extensive rifting in marginal basins

NO STABLE CRATONS

The roots of cratons are chemically buoyant mantle material that is stiff and resistant, contributing to their survival from the Archean to the present *(Carlson et* al.2005).

Subduction Vs Plume origin of Archean Mantle Lithosphere

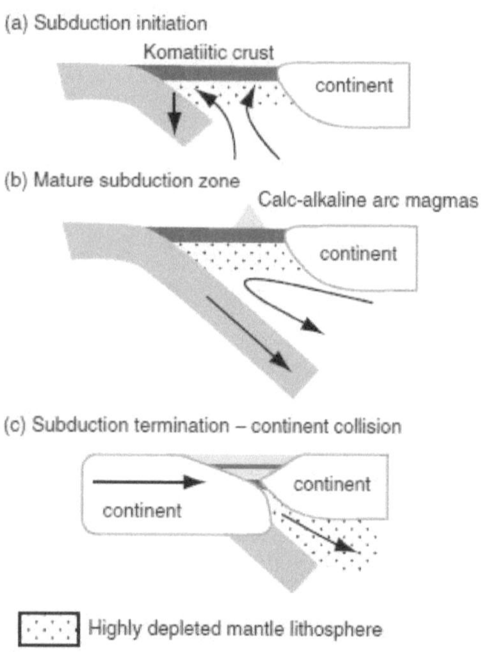

Arid source rocks:

(a) Partial melting - forearc setting

(b) Cool and hydrated residual mantle

(c) Obduction of komatiitic crust during collision

Fig.4 Conceptual model for the generation of komatiites and cratonic mantle by partial melting in a subduction zone (*after Parman et al., 2004. Copyright © 2004 Geological Society of South Africa).*

Subduction Vs Plume origin of Archean Mantle Lithosphere

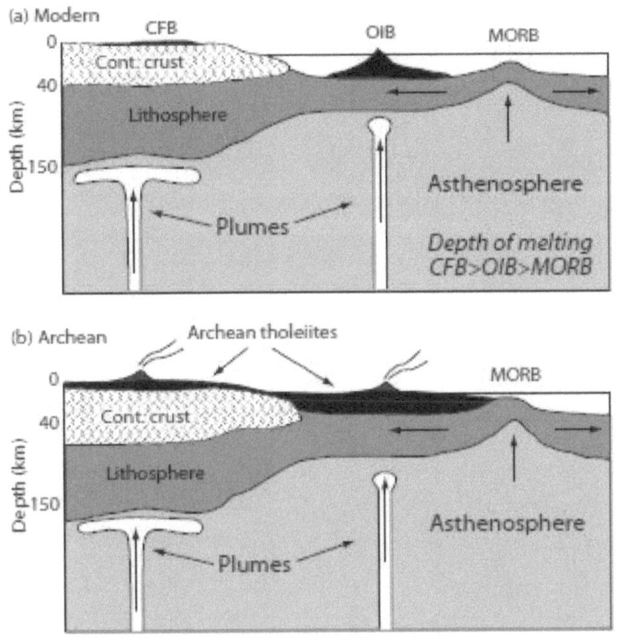

Fig. 5 Model of komatiitic and tholeiitic basalt formation involving mantle plumes (after Arndt et al., 1997, by permission of Oxford University Press). Model shows the influence of lithospheric thickness on depth of melting where CFB is continental flood basalt, OIB oceanic island basalt, and MORB mid-ocean ridge basalt.

Dry rock source:

Melting begins at much greater depths:

- Mantle plumes
- Deep levels beneath mid-ocean ridges.

Crustal structure – Special Dome-and-Keel Architecture

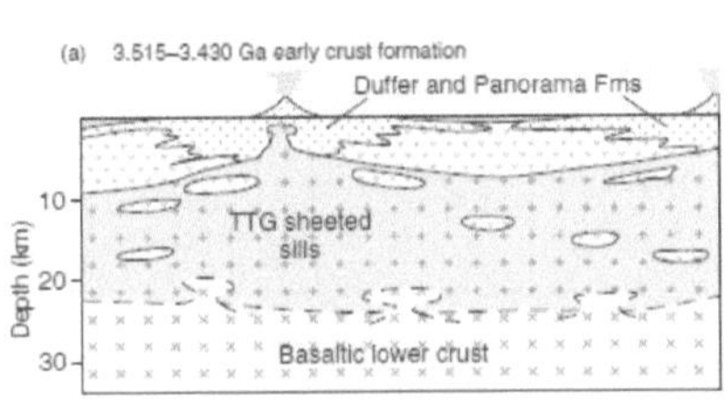

Fig. 6 Three-stage diapiric model of dome-and-keel provinces in the Eastern Pilbara craton (after Collins et al., 1998, with permission from Pergamon Press, Copyright Elsevier 1998; the age-spans of the stages are from Van Kranendonk et al., 2007).

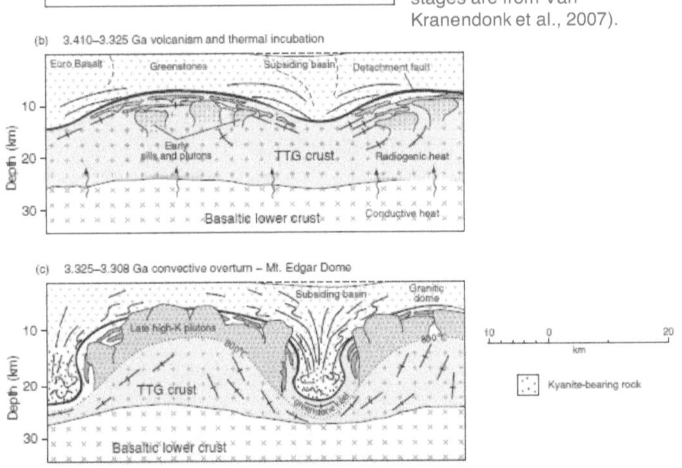

Vertical tectonic models :

- Diapiric rise of hot granitoid domes
- Burring and partial melt
- Sinking of greenstones – rising of partial melt

Horizontal tectonic models *:*

One or more periods of,

- Contraction – Collision and Terrain accretion
- Extension – Detachment zones – emplacement of granitoid domes

Both have their constrains but both tectonics had important role

Archean – Proterozoic Boundary

TECTONIC INFER

From Vertical movements towards Horizontal

Obvious driving mechanism – Secular cooling of the Earth

"Protocrust" Formation

Late Archean thickened continental crust:
- Uplift and erosion
- First basins covered by clastic debris

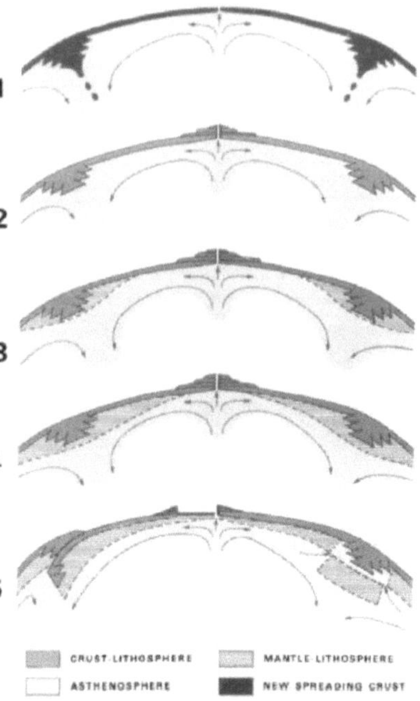

Fig. 7 Pictorial cartoon illustrating postulated evolution of tectonic style. R. B. Hargraves. 1981. PRECAMBRIAN PLATE TECTONICS. Ch. 2 Precambrian tectonic style: a liberal uniformitarian interpretation, p.46

Proterozoic Eon 2,5-0,6 Ga

- Rigid Plates – STABLE CRATONS
- Some greenstone belts continue form (Early Proterozoic)
- Continental break up and rearrangement
- Wilson Cycle recognized (subduction, collision and terrain accretion)
- Progressive change of bulk composition of the crust *(Condie, 2005b)*
- Resisting of the large cratons to large-scale lithospheric recycle
- Crust isolation from deeper convecting Mantle

Crust Production:
Episodic with rapid growth at 2,7, 1,9 and 1,2Ga and slower growth afterwards

Most important Early Proterozoic Continental Crust growth mechanisms:

- Magma addition *(Condie, 1998; Wyman & Kerrich, 2002)*
- Terrane accretion *(Pearson et al., 2002)*

Typical Wilson Cycle Tectonics After Middle Proterozoic

FIRST OROGENS – Lithospheric plate margins

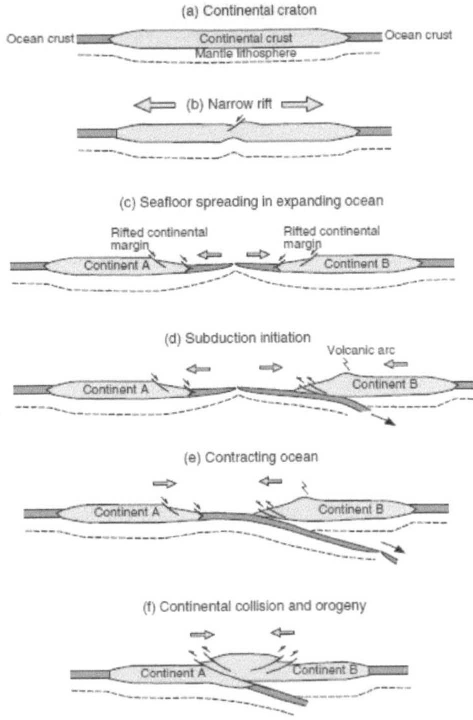

Fig. 8 The Wilson cycle. Kearey et. Al..2008. Global Tectonics. Chapter 7, p.208

Supercontinental Cycle (Late Archean – Early Proterozoic):

a) Single Supercontinent Model *(Williams et. Al., 1991)*
b) Three Supercratons Model *(Bleeker, 2003)*
c) Many small dispersed landmasses Model *(Bleeker, 2003)*

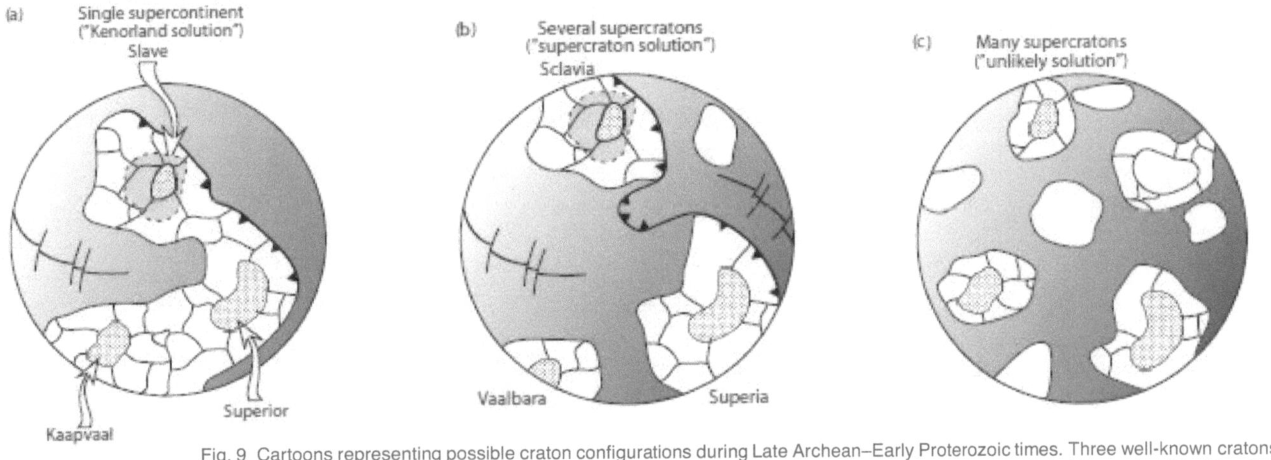

Fig. 9 Cartoons representing possible craton configurations during Late Archean–Early Proterozoic times. Three well-known cratons (Slave, Superior and Kaapvaal) (after Bleeker, 2003, with permission from Elsevier). Kearey et. Al..2008. Global Tectonics. Ch. 11, p.374

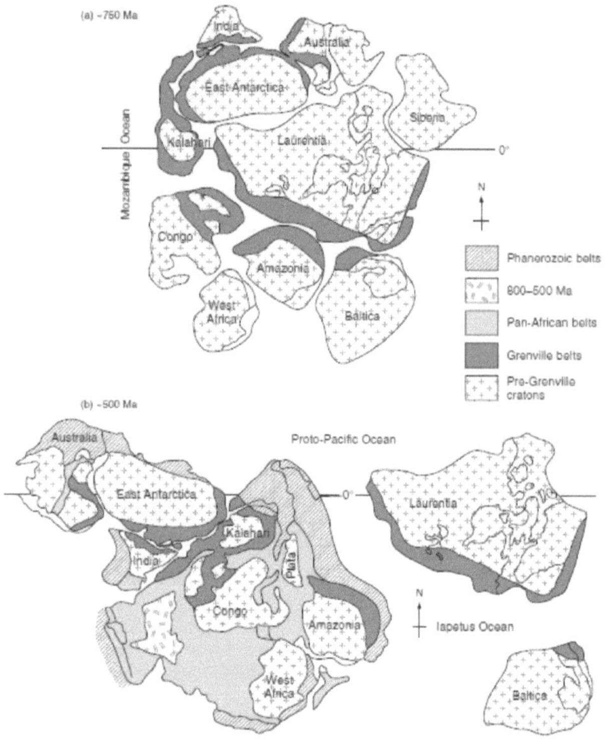

Late Proterozoic Supercontinent – Rodinia

- Supercontinent fragment
- East Antarctica, Australia and India blocks anticlockwise movement
- Opening of proto-Pacific Ocean
- Congo, West Africa and Amazonia block clockwise movement ?
- Block Collision -> Gondwana
- Gondwana clockwise rotation away from Laurentia
- Iapetus Ocean opening
- Iapetus Ocean closing -> Pangea 250Ma

Fig. 10 (a) Reconstruction of the Late Proterozoic supercontinent Rodinia. (b) Late Cambrian paleogeography after the break-up of Rodinia and the formation of Gondwana (after Hoffman, 1991, with permission from Science 252, 1409–12 with permission from the AAAS). Kearey et. Al..2008. Global Tectonics. Ch. 11, p.371

Use of Paleomagnetic data: Pannotia Reconstruction *(Dalziel, 1997)*

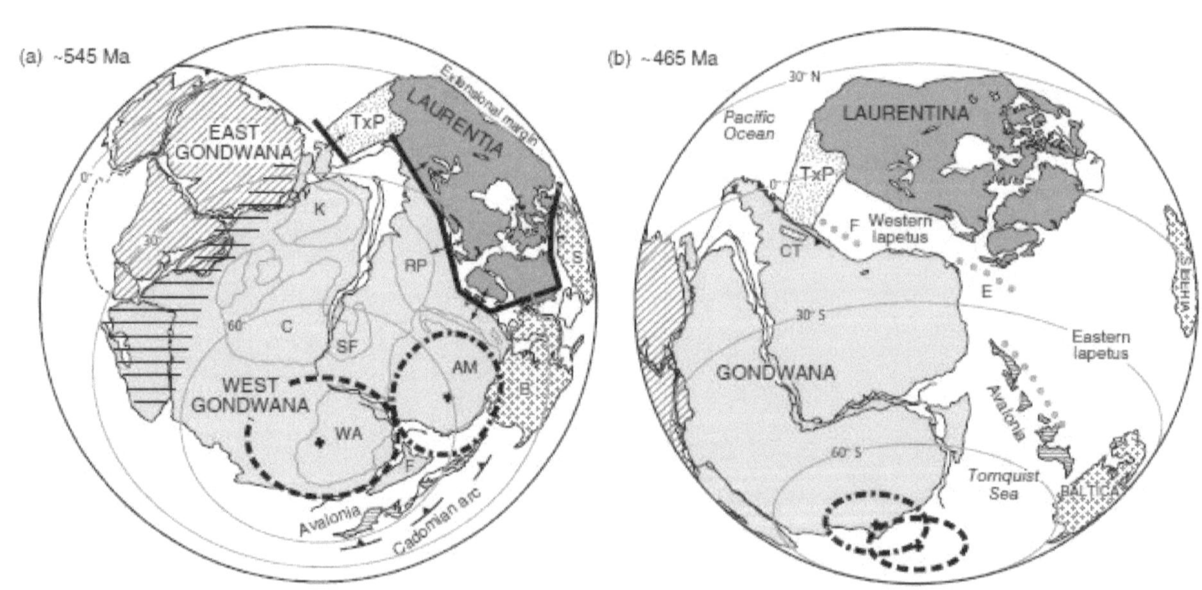

Fig. 11 Postulated reconstructions of (a) Pannotia at ~545 Ma and (b) the rifting of Laurentia and Gondwana at ~465 Ma emphasizing the paleogeography of Laurentia relative to Gondwana (after Dalziel, 1997, with permission from the Geological Society of America) Kearey et. Al..2008. Global Tectonics. Ch. 11, p.3375

Introduction	Hadean Eon (4,5-3,9 Ga)	Archean Eon (3,9-2,5 Ga)	Archean – Proterozoic Boundary	Proterozoic Eon (2,5-0,6 Ga)	Super - Continental Cycle	Summary

Summary

Hadean Eon 4500-3900 Ga	Controversial - Formation of proto-crust and proto-lithosphere of non successfully determined composition. No horizontal tectonics.
Archean Eon 3900-2700 Ga	High rate of radiogenic decay, small-scale mantle convection and vigorous high formation-destruction rates of oceanic crust which was subducted. Extensive rifting in marginal basins.
Archean - Proterozoic Boundary 2700-2300 Ga	Transitional and diachronous period during which much uplift and erosion of late Archaean thick continental crust. Stabilization of crust-lithosphere following aggregation of Archaean proto-plates. Many dykes and stratiform intrusions injected into stabilizing crust.
Proterozoic Eon 2300-600 Ga	Lower mantle temperatures and slower global spreading rate leads to modern Wilson-cycle from early Proterozoic. Narrow geosynclines and mobile belts giving rise to extensive more stable plates. First Supercontinents.
Phanerozoic Eon 600 Ma to Present	Wilson cycle controls all geological-tectonic phenomena

Alternated Table by Anastasia Kokkinou after Brian F. Widdley. A. Kroener. PRECAMBRIAN PLATE TECTONICS.1981. Chapter 1, p.16

Bibliography

A. Kroener (Editor) et. Al.. 1981. Developments in Precambrian Geology 4 - PRECAMBRIAN PLATE TECTONICS. ELSEVIER SCIENTIFIC PUBLISHING COMPANY. Institut fcr Geowissenschaften, Johannes Gutenberg-Universitgt Mainz, Mainz, Federal Republic of Germany. Chapters 1, 3, 4, 18, 23 & 26.

Tracy Rushmer, John Adam, Jonathan O'Neil and Don Francis. 2013. Hadean greenstones from the Nuvvuagittuq fold belt and the origin of the Earth's early continental crust. © 2012 Geological Society of America, Geology 2012;40;363-366, doi: 10.1130/G32623.1. 1GEMOC, Earth and Planetary Sciences, Macquarie University, NSW 2109, Australia, Department of Terrestrial Magmatism, Carnegie Institution of Washington, 5241 Broad Branch Road N.W., Washington, D.C. 20015, USA, Earth and Planetary Sciences Department, McGill University, 3450 University Street, Montreal, Quebec H3A 2A7, Canada

Tracy Rushmer, Simon Turner, Mark Reagan, Jean-Francois Moyen. 2015(Not yet published). Hadean Subduction. Department of Earth and Planetary Sciences, Macquarie University, Sydney, Australia, Department of Geoscience, University of Iowa, Iowa City, Iowa 52242, USA UMR 6524 CNRS and Universite Jean-Monnet, 23 rue du Dr. Michelon, Saint-Etienne, France.

Philip Kearey, Keith A. Klepeis, Frederick J. Vine. 2008. Global Tectonics. John Wiley & Sons Ltd Publications UK. Chapter 11 – Precambrian Tectonics.

Thank You for Your Attention